Hertzian Tales

Hertzian Tales

Electronic Products, Aesthetic Experience, and Critical Design

Anthony Dunne

The MIT Press

Cambridge, Massachusetts

London, England

MIT Press books may be purchased at special quantity discounts for business or sales promotional use. For information, please email special_sales@mitpress.mit.edu or write to Special Sales Department, The MIT Press, 55 Hayward Street, Cambridge, MA 02142.

This book was set in Bell Gothic and Garamond 3 by Graphic Composition, Inc. Printed and bound in the United States of America.

Library of Congress Cataloging-in-Publication Data

Dunne, Anthony.
Hertzian tales : electronic products, aesthetic experience, and critical design / Anthony Dunne.—[Rev. ed.]
  p. cm.
Includes bibliographical references and index.
ISBN 0-262-04232-0 (hc : alk. paper)
1. Electronic apparatus and appliances—Design and construction. I. Title.

TK7870 D86 2006   621.381—dc22   2005052222

10 9 8 7 6 5 4 3 2 1

Contents

Foreword to the 1999 Edition

Initially funded by a generous grant from the Silicon Valley company Interval Research Corporation, the Computer Related Design Research Studio at London's Royal College of Art was founded to investigate how the skills and knowledge of artists and designers might be applied to the design of information technology systems and products. Bringing together a variety of perspectives—architecture, industrial and graphic design, psychology, engineering, and fine art—the studio now collaborates on projects with companies and organizations worldwide. This book is the first of several bringing the work of the studio to a wider audience.

In the 1950s Reyner Banham, in *Theory and Design in the First Machine Age,* was already describing the arrival of the second, "the age of domestic electronics and synthetic chemistry." If the First Machine Age was heroic (cars, planes, heavy industry, speed), in the second, the age of small machines (vacuum cleaners, washing machines, electric mixers) made possible by electricity and advances in control technology, the machine was literally domesticated, an ingredient of everyday experience. Forty years later, we live in the Third Machine Age, an age of electronic devices that hardly seem like machines at all. Made possible first by the transistor and later by the silicon chip, computers were initially designed by technologists as technical tools for other technologists. They then entered the office. Only recently has computer technology, converging with telecommunications technology, invaded the home.

The work of Dr. Anthony Dunne, who trained and has practiced as an industrial designer, is innovative, haunting, but always benevolent and clear-sighted.

In *Hertzian Tales,* he reflects on the historical and conceptual context of this work in the light of the electronic infiltration of everyday life.

This vast potential of "electronic omnipresence" has hardly begun to be recognized. Manufacturers tell us our toaster is "electronically controlled," but it still just toasts bread. Our washing machine may be "electronic," but, though with less human effort, it does no more than the washboard and wringer. Yet electronic machines could serve altogether new purposes. If the first two machine ages were truly mechanical extensions of our musculature, electronic machines can extend our minds, our subjective relationship with the world, our active relationship with other people.

Back in the Second Machine Age, in 1951, my family bought its first refrigerator—a gas one, because it was cheaper to run. Our previous means of delaying food decay had been earthenware coolers soaked in water, marble slabs, and leaving food outside in winter. Our new fridge kept things cold for longer. It also signified for us, however, the end of post-war austerity and the family's progressive optimism. Objects of use, in short, seldom have solely practical significance, but also carry ritual and symbolic meaning. This has long been understood by manufacturers of consumer products (our fridge was vaguely streamlined to withstand the hurricane pace of our newfound modernity). But in the design of computer technology, this aspect has until recently been either considered simplistically, as a matter of "executive" styling, for instance, or ignored.

Objects change not just how we see ourselves but, moreover, how we live life. The domestic fridge reduced the need for food to be bought daily and cooked twice daily. Partly in consequence, female employment patterns changed and family life began to lose its ritual ceremony of solidarity, hierarchy, and manners: the shared mealtime.

To suggest how electronic objects might enrich rather than impoverish the lives we now share with them, Dunne proposes an "aesthetics of use": an aesthetics that, through the interactivity made possible by computing, seeks a developing and more nuanced cooperation with the object—a cooperation that, it is hoped, might enhance social contact and everyday experience. Such an aesthetics, clearly, attends less to how an object looks, the traditional concern of product aesthetics, than to how it behaves.

Not surprisingly, the development of new electronic products is currently led by technology companies and what they think will sell. To reduce risk, this development tends to be incremental: customers will probably like a product similar to a predecessor that has already sold well, and its manufacture should present few new problems. But this conservatism, though understandable, tends only to offer ever more intelligent toasters. Architecture has a long tradition of the "ideas competition" whose entries are not usually intended to be built but to publicly disseminate radical ideas about how architecture, and possibly the life it accommodates, might be differently conceived. The world of electronic product design needs a similar speculative arena, to imagine possible and impossible futures with computer technology, and to extend what people might find enjoyable to live with and what manufacturers could imagine making and selling. The Computer Related Design Research Studio aims to be part of this arena. Anthony Dunne and Fiona Raby, founding members of the studio, contribute to it their distinctive and powerful voices.

Hertzian Tales explores a space between fine art and design, showing how designers can use fine-art means—provoking, making ambiguous, making strange—to question how we cohabit with electronic technology and to probe its aesthetic potential. Dunne describes previous approaches to the design of electronic objects, and, more generally, how other disciplines—semiology, anthropology, design, and literature—have conceived the cultural role of the object. He then considers the "aesthetics of use" and what he calls "psychosocial narratives" and "real fictions," concentrating less on sculpture, the traditional reference for industrial designers, and more on the "complicated pleasures" of literature and film.

Dunne believes that "the most difficult challenges for designers of electronic objects now lie not in technical and semiotic functionality, where optimal levels of performance are already attainable, but in the realms of metaphysics, poetry, and aesthetics, where little research has been carried out." In these realms, he is a bold pioneer.

Gillian Crampton Smith
Professor of Computer Related Design
Royal College of Art
1998

Preface to the 2005 Edition

The ideas in *Hertzian Tales* were developed between 1994 and 1997 while I was completing my Ph.D. thesis in the Computer Related Design department at the Royal College of Art in London. The first edition was published through the Royal College of Art in 1999.

It is interesting to look back and think about the technological developments made since then. Bluetooth, 3G phones, and wi-fi are all now part of everyday life. The dot-com boom has come and gone. And in the United Kingdom, large parts of the electromagnetic spectrum are about to be deregulated.

Yet very little has changed in the world of design.

Electronic technologies are still dealt with on a purely aesthetic level. There are some exceptions, of course: Droog's Bootleg range of modified classic electronic products, Ron Arad's prototype tables with surfaces that act as low-resolution screens, his textable chandeliers for Swarovski, and of course Apple's iPod. But still, something is missing. Design is not engaging with the social, cultural, and ethical implications of the technologies it makes so sexy and consumable.

Today, new biotechnology products and services are announced almost daily, while newspaper stories regularly explore the ethics of genetic identity, DNA theft, biopiracy, designer babies, consumer eugenics, genetic underclass, and molecular surveillance.

Nanotechnology is no longer science fiction. Insurance companies are already anticipating its consequences. Policies are being developed to cover workers in new labs springing up around the world who might accidentally inhale nanoparticles.

Although the technological focus of this book is electronics, I hope that its main argument, that design can be used as a critical medium for reflecting on the cultural, social, and ethical impact of technology, is even more relevant today.

Anthony Dunne
December 2004

Acknowledgments

I would like to thank the following for their help throughout this research project: Kei'ichi Irie, John Thackara, and Daniel Weil for their encouragement and advice while getting the project under way; Gillian Crampton Smith for her supervision and encouragement during the development of the Ph.D. thesis; Phil Tabor for his invaluable critical observations and advice during the final drafts of the thesis; Alex Seago for his helpful comments on early drafts; CRD staff and students for thorough discussion and debate during the development of this project; the many people who provided opportunities to test the ideas in this book through lectures, workshops, and teaching; Spencer Childs for technical advice about the world of radio; Jayne Roderick for sound design; Dan Sellars for the *Pillow Talk* video; Lubna Hammoud for photographic portraits of Thief of Affections; Shona Kitchen for detective work and picture research; Giles Lane for his editing of the first edition; and Fiona Raby, my partner and collaborator, for inspiration, reality checks, and endless support.

I would also like to thank Douglas Sery and Kathleen Caruso at The MIT Press, and Nina Due for invaluable help in preparing this edition.

Introduction

Whereas architecture and furniture design have successfully operated in the realm of cultural speculation for some time, product design's strong ties to the marketplace have left little room for speculation on the cultural function of electronic products. As ever more of our everyday social and cultural experiences are mediated by electronic products, designers need to develop ways of exploring how this electronic mediation might enrich people's everyday lives.

Hertzian Tales explores the way critical responses to the ideological nature of design[1] can inform the development of aesthetic possibilities for electronic products. It focuses on the role they play in shaping our experience of inhabiting the "electrosphere," looking beyond the quality of our relationship with objects themselves to the aesthetics of the social, psychological, and cultural experiences they mediate.

The primary purpose of this book is to set the scene for relocating the electronic product beyond a culture of relentless innovation for its own sake, based simply on what is technologically possible and semiologically consumable, to a broader context of critical thinking about its aesthetic role in everyday life.

The project proposes an approach that uses the design of conceptual electronic products as a way of provoking complex and meaningful reflection on the ubiquitous, dematerializing, and intelligent artificial environment we inhabit.

The origin of *Hertzian Tales* comes out of two places: first, a frustration with the limited role currently played by industrial designers (compared to those of engineers and marketers) in the development of new electronic products; and, second, a belief that design, too, has much to contribute as a form of social commentary, stimulating discussion and debate among designers, industry, and the public about the quality of our electronically mediated life.

The book aims to map out the current technologically informed environment of ideas about electronic objects and to understand industrial design's potential within it—developing a position that is both critical and optimistic.

Industrial design is not art, but neither is it purely a business tool. While mainstream industrial design is comfortable using its powerful visualization capabilities to propagandize desires and needs designed by others, thereby maintaining a society of passive consumers, design research in the aesthetic and cultural realm should draw attention to how products limit our experiences and expose to criticism and discussion their hidden social and psychological mechanisms.

I believe strongly in the potential of industrial design as applied art, or industrial art, to improve the quality of our relationship to the artificial environment, and in industrial design's potential, at the heart of consumer culture, to be subverted for more socially beneficial ends. In order to achieve this, research is needed into an expanded notion of design aesthetics that includes more poetic and metaphysical relationships with the artificial environment of technological artifacts.

Although much has been written about the interrelationship of technology and culture, few sources address design specifically—and there is a surprising lack of examples of such research within industrial design. Even the cultural and aesthetic experiments of design groups like Memphis, or more recently Droog Design, rarely touch on electronics (although arguably the general changes brought about through electronic technologies have played a part in motivating such work). Because of this, this book draws many examples from other fields such as fine art and architecture.

Not surprisingly, research such as this is difficult to situate in a purely commercial context. Academia provides a space for developing ideas and approaches—"hypotheses for action"—but, as in technical research, an industrial context better suits the complexities of realization and implementation. An academic setting can also provide a new context for design outside commercial practice, distinct from the more established critical approaches of architecture and fine art.

Andrea Branzi and other experimental designers of the 1960s and 1970s addressed the role played by design in poetic modes of inhabitation and, guided by an architectural perspective, developed provocative research approaches and positions focusing on the expressive and linguistic possibilities of new materials and surfaces.[2] During the early 1980s, Daniel Weil produced a group of design proposals for radios, applying some of the concerns of Italian radical design to electronic products. More recently, Ezio Manzini outlined a role for the designer that offers a fresh perspective that builds on earlier Italian design thinking. He suggests that the days of the design visionary are over, and a weariness with utopian visions has set in. Instead, he advises the designer to use his or her skills to visualize alternative future scenarios in ways that can be presented to the public, thus enabling democratic choices between the futures people actually want. Designers could then set about achieving these futures by developing new design strategies to direct industry to work with society.

But in the *Hertzian Tales* project, the designer, like J. G. Ballard's writer,[3] no longer knows anything for certain; all he or she can offer are the contents of his or her own head, where internal imagination meets the external world of reality. Design is used as a strategy for linking these two worlds. Its outcome consists of conceptual design proposals offering a critique of the present through the material embodiment of functions derived from alternative value systems. These "material tales" are not utopian visions or blueprints—clear-cut modeling of the future is too didactic. Instead, they mix criticism with optimism to provide the "complicated pleasure"[4] found in other imaginative media such as film and literature, particularly those that explore boundaries between the real and the unreal.

Hertzian Tales consists of two main parts: six essays exploring design approaches for developing the aesthetic and critical possibilities of electronic products outside a commercial context (chapters 1–6), and five conceptual design proposals expressed as objects, videos, and images—by-products of an investigation into a synthesis between practice and theory, where neither practice nor theory leads (chapter 7).

Chapter 1, "The Electronic as Post-optimal Object," discusses existing design approaches to the development of aesthetic possibilities for electronic objects. Current design approaches aim to optimize the experience of using an object, with the effect of constraining our experience to the prosaic. However, when practicality and functionality can be taken for granted, the aesthetics of the "post-optimal" object provide a much richer field of investigation.

If user-friendliness characterizes the relationship between people and the optimal electronic object, then user-unfriendliness, a form of gentle provocation, could characterize the post-optimal object. "(In)human Factors" (chapter

2) discusses a role for different degrees of user-unfriendliness such as estrangement and alienation in the design of electronic objects.

Design's emphasis on forms of expression and languages of representation, rather than on experience, holds back the potential of electronic objects to provide new levels of aesthetic experience. "Para-functionality" (chapter 3) investigates the design of function (rather than form) to provide new types of aesthetic experience.

"Psychosocial Narratives" (chapter 4) discusses behavior as a narrative experience determined by objects, and considers how the embodiment of unusual psychological needs and desires in electronic objects can encourage the user to experience new narrative situations as a protagonist. These new possibilities complement normality by referring to the world of object misuse and abuse.

"Real Fiction" (chapter 5) discusses systems of presentation and consumption for ideas that, unlikely to be mass-produced or even prototyped, exploit the conceptual status of objects as ideas.

"Hertzian Space" (chapter 6) explores how artists and designers have made links between the invisible environment of electromagnetic radiation and the material culture of objects, and draws attention to the neglected aesthetic dimensions of electromagnetic fields.

"Hertzian Tales and Sublime Gadgets" (chapter 7) is a commentary on the design work for this project. It consists of five conceptual design proposals for the electronic as a post-optimal object: Electroclimates (with Fiona Raby), When Objects Dream . . . , Thief of Affections, Tuneable Cities (with Fiona Raby), and Faraday Chair. They are not necessarily illustrations of the ideas discussed in earlier chapters, nor are the earlier chapters an explanation of these proposals. They evolved simultaneously and are part of the same design process.

1

The Electronic as Post-optimal Object

As new technical developments alter the object and make it "intelligent," they also set the object on a plane with no prior cultural references . . . although the physical aspects of these objects are still within the world of materials, their operation and their very state of being is well beyond the manipulation of matter and has more to do with information exchange than with form.
—E. MANZINI, *THE MATERIAL OF INVENTION*

Most designers of electronic objects have responded to this challenge by accepting a role as a semiotician, a companion of packaging designers and marketers, creating semiotic skins for incomprehensible technologies.

From Banham writing about portable radios in the 1970s, through the plethora of essays on "product semantics" in the 1980s, to Norman Bolz's 1992 essay "The Meaning of Surface," the treatment of the electronic object as a package for technology, designed to communicate use, cultural meaning, and corporate identity through its surface, has been thoroughly explored. The electronic object accordingly occupies a strange place in the world of material culture, closer to washing powder and cough mixture than to furniture and architecture, and is subject to the same linguistic discipline as all package design, that of the sign. It is lost somewhere between image and object, and its cultural identity is defined in relation to technological functionalism and semiotics.

This chapter considers three perspectives on the electronic object: "The Electronic as Lost Object" briefly discusses theoretical perspectives, "The Electronic as Object" focuses on design approaches, and "The Electronic as Post-optimal Object" introduces the idea of the "post-optimal" object.[1]

The Electronic as Lost Object

A Technological Perspective

From a technological perspective the theories of Jean Baudrillard and Paul Virilio are a stimulating source of ideas about the effects of electronic technology on the way we experience and think about ourselves, objects, and environments. Their provocative fusions of analysis and imagery offer a rich inspiration while remaining grounded in reality. But there is a danger that if designers are seduced by this, their designs will become mere illustrations of descriptions of electronic objects. Designers of electronic objects are already familiar with the kinds of technologies analyzed by these writers. It is more important to extend the range of cultural values, building on what is already understood, rather than illustrating it.

Some writers on the social history of technology present the ideological dimension of everyday technologies, even if these are often pre-electronic. This is useful to critique the human factors "community," who have developed a view of the electronic object, derived from computer science and cognitive psychology, that is extremely influential in the computer industry; see, for example, Don Norman's (1988) *The Psychology of Everyday Things.*

A serious problem with the human factors approach though, in relation to this project, is its uncritical acceptance of what has been called by Bernard Waites (1989) the "American Ideology," or the ideological legitimation of technology:

All problems whether of nature, human nature, or culture, are seen as "technical" problems capable of rational solution through the accumulation of objective knowledge, in the form of neutral or value-free observations and correlations, and the application of that knowledge in procedures arrived at by trial and error, the value of which is to be judged by how well they fulfil their appointed ends. These ends are ultimately linked with the maximisation of society's productivity and the most economic use of its resources, so that technology, in the American Ideology, becomes "instrumental rationality" incarnate, the tools of technocracy. (31)

The result, as the computer industry merges with other industries, is that the optimization of the psychological fit between people and electronic technology, for which the industry strives, is spreading beyond the work environment to areas such as the home that have so far acted as a counterpoint to the harsh functionality of the workplace. When used in the home to mediate social relations, the conceptual models of efficient communication embodied in office equipment leave little room for the nuances and quirks on which communication outside the workplace relies so heavily.

Writing on electronic art might seem a good source of ideas on the electronic object, but, surprisingly, electronic art has become so technology-driven that it seems concerned only with the aesthetic expression of technology for its own sake. Rather than relating the impact of technology to everyday life, art criticism in this area glamorizes technology as a source of aesthetic effect to be experienced in galleries. The exceptions tend to be based on electronic systems rather than objects (e.g., in the work of Roy Ascott).

A Semiotic Perspective

A semiotic approach has been taken by design writers, both at the linguistic level, looking at the way objects can be "written" and "read" as visual signs, and at the more general level of the study of consumerism, where semiological analysis of objects as commodities has revealed their part in maintaining what Roland Barthes (1989) has called "mythologies." An impressive semiotic analysis of the object is Baudrillard's (1981) *For a Critique of the Political Economy of the Sign,* which shifts the emphasis on the analysis of commodities away from the production of objects to the consumption of signs.[2] But, as Daniel Miller (1987) writes: "While the rise of semiotics in the 1960s was advantages [sic] in that it provided for the extension of linguistic research into other domains, any of which could be treated as a semiotic system, this extension took place at the expense of subordinating the object qualities of things to their word-like properties" (95–96).

A Material Culture Perspective

Although there is very little available on the electronic object, the study of material culture is still of interest because it situates the object firmly within everyday life. Academically, it is somewhere between anthropology, sociology, and ethnology.

Miller (1987) claims there is a an "extraordinary lack of academic discussion pertaining to artifacts as objects, despite their pervasive presence as the context for modern life" (85), and provides an alternative to the semiological analysis of mass consumption by distinguishing material culture from language and the study of meaning in order to focus on the physical nature of artifacts.

In contrast to analyses of the object in relation to consumerism, Mihaly Csikszentmihalyi and Eugene Roshberg-Halton's (1981) *The Meaning of Things* analyzes the meaning of objects in domestic settings, emphasizing their symbolic role. And in "The Metafunctional and Dysfunctional System: Gadgets and Robots," Baudrillard (1996) writes about the electronic gadget as the subject of a science of imaginary technical solutions. Although originally written nearly thirty years earlier, Baudrillard's analysis of electronic gadgets is far more stimulating than a more recent analysis, *Consuming Technologies,* by Roger Silverstone and Eric Hirsch (1992), which is more concerned with descriptive models than with Baudrillard's challenges to the imagination. *Hertzian Tales* is more concerned with "critical" theories,[3] and thus in assessing the development of objects not against whether they fit into how things are now, but the desirability of the changes they encourage.

The value of material culture for this study is that it draws attention to the complex nature of our relationship to ordinary objects and provides standards against which new electronic objects can be compared.

A Design Perspective

Since the early 1960s a very narrow form of semiotic analysis has dominated design thinking about the electronic object. Of books written about design from a theoretical point of view, only John Thackara's (1988) *Design after Modernism* contains new perspectives on the electronic object.

Books and articles by designers, based on particular projects prove more interesting. Manzini and Susani (1995) present a collection of design projects that explore a place for solidity within the fluid world created by electronic technology: "In the fluid world the permanent features we need are no longer there as a matter of course, but are the result of our desire; the 'solid side' in a fluid world, if and when it exists, will be the result of a design" (16).

Their strong emphasis on aesthetics and ecological concerns is a powerful example of design research carried out by practicing designers within an intellectual context. Susani has developed a design perspective that locates the electronic object within material culture rather than semiology or electronic media. He writes: "We are lacking a discipline, perhaps an 'objectology,' or an 'object ethology,' which allows us to analyse and systematise objects and to formulate the rules and codes of their behaviour . . . a discipline which recovers and updates the interrupted discourse of material culture, in crisis since the world of objects was taken over by the world of products and the world of consumption" (Susani 1992, 42). He also recommends a sensual approach to introducing technology into the home, building on what is already there, and exploring the overlap between the material and immaterial world from an aesthetic and anthropological point of view. He suggests that material culture could offer useful insights to this problem.

A Literary Perspective

However, the most fruitful reflection on material culture is to be found, not in anthropology or sociology, but in literature concerned with the poetry of everyday objects. In *The Poetics of Space,* Gaston Bachelard (1969) offers an analysis, influenced by psychoanalysis, that emphasises the poetic dimension of humble furniture such as wardrobes and chests of drawers; Jun'ichiro Tanizaki's (1991) *In Praise of Shadows* considers the Japanese object in relation to shadows and darkness, and the effects of electricity on their appreciation; and Nicholson Baker's novels (such as *The Mezzanine* (1986) and *Room Temperature* (1990)) give everyday industrial products significant roles.

The view of objects suggested by literary writers reveals a poetry of material culture that offers a fresh alternative to the formal aesthetic criticism of the art object and to the academic analysis of their meaning as signs. Their objects are firmly grounded in everyday life.

The best writing in this area blends anthropology, sociology, and semiology to explore the irrational dimensions of the material culture of everyday life. As the electronic object rarely features in this literature, the discussion in the rest of this book is based mainly on design proposals.

The Electronic as Object

This section discusses four design approaches to the electronic object: packages, fusions, dematerialization, and juxtaposition. They differ in how each addresses the conflict between the solidity of the object and the fluidity of electronic media. Design is viewed here as a strategy for linking the immaterial and the material.

Packages

Commercial design's approach to the electronic object has been to treat it as a package for electronic technology. An example of this, where the aesthetic and conceptual possibilities of the package are thoroughly exploited, is Daniel Weil's *Radio in a Bag* (figure 1.1), which takes the idea of the designer's role as a packager of technology to the extreme. On one level the electronics provide decoration, while on another, their exposure signals a nonchalance toward technology. The radio's literal flexibility expresses the flexible structural relationship between electronic components, and its transparency attempts to demystify the electronic

Figure 1.1 Daniel Weil's *Radio in a Bag* (1983) takes the idea of the designer's role as a packager of technology to the extreme.

object. It shows that by taking a playful approach to package design and liberating it from product semantics, even the packaging of electronics can yield interesting results. Ironically, part of the critical success of this design, despite being a package, is its treatment as a thing rather than an image.

Fusion

The logic of computers is expressed in forces that are averages of the behaviour of many electrons. No machine has ever been so far removed from the world of human experience: the largest aircraft carriers are still infinitely closer to the human scale than the simplest, slowest microcomputers.
—D. J. BOLTER, *TURING'S MAN*

The electronic object is a confusion of conceptual models, symbolic logic, algorithms, software, electrons, and matter. The gap between the scales of electrons and objects is most difficult to grasp.

The architect Neil Denari has spoken of the need for the "overcoming of the symbolic," and his view is that architecture must make a connection between the worlds of electromagnetism and spatial inhabitation. But there is greater chance of bridging the gap between electromagnetism and inhabitable space if one where to explore this route through the design of objects rather than buildings.

The first transistor (figure 1.2) is a test-rig for a key electronic component created by inventors who work at the level of both electrons and matter. They organize matter as interacting volumes of electrons,[4] and they offer a possibility for reconciling the scales that separate the worlds of electrons and space. But once these prototype elements have been subjected to the extreme rationalization required by mass production, they become reduced to abstract ultra-miniaturized electronic components. Their modernist poetry, based on truth to materials, is lost.

Closing the gap between the scales of electronics and objects by directly manipulating materials as volumes of electrons is a difficult route for designers. This task is essentially limited to scientists, and even their test-rigs will eventually become miniaturized components. *Clock* by Daniel Weil (figure 1.3) captures some of this quality—partly a reaction against miniaturization, its size is based on the largest circuit boards available in the early 1980s. The circuit is composed visually and the wires linking the two main components are made from dining forks. Familiar objects are put into new but natural relationships based on electrical properties.

Figure 1.2   The first transistor: a test-rig for a key electronic component created by inventors working at the level of both electrons and matter.

This approach resembles the way electricity was dealt with in early natural philosophy books that explained electricity in delightfully poetic ways, drawing attention to unusual but real phenomena: "The simultaneous development of both kinds of electricity is illustrated by the following experiment:—Two persons stand on stools with glass legs, and one of them strikes the other with a catskin. Both of them are now found to be electrified, the striker positively, and the person struck negatively, and from both of them sparks may be drawn by presenting the knuckle" (Everett, *Deschanel's Natural Philosophy,* Part 3, 4).

The development of "smart materials" is another area where the gap between the electronic and material is being closed, although primarily for technical reasons. Scientists and engineers are developing new materials, designed at a molecular level, that are responsive, dynamic, and almost biological. Although most of these materials are still experimental some, such as electroluminescent laminates[5] and piezoelectric films[6] have been around for several decades.

Manzini (1986) explores the implications of designing with these new smart materials: "The design of this skin, and therefore of the objects that are made with it, is chiefly the design of interactivity with the environment—a scenario for which we must prepare the stage, the sets, and the actors. Imagining the nature of these 'individual objects' is another new chapter in the history of design" (204). Most of Manzini's specially commissioned examples illustrate the miniaturization arising from integrating previously separate mechanisms and their novel decorative possibilities. However, they do not demonstrate the radical aesthetic potential of these materials to open new channels of communication with the environment of electronic objects.

Only Alberto Meda and Denis Santachiara's *Stroke Lamp* (see Manzini 1986) hints at the new relationships between people and machines made possible through new reactive materials. It is controlled by stroking the surface, which is made from an insulating plastic with a copper circuit deposited on it by a photochemical process similar to that used for printed circuits. Although low-tech, it suggests a sensual and playful interaction with everyday objects that might be extended to more complex interactions as more sophisticated materials become available. Andrea Branzi's *Leaf Electroluminescent Lamp* (1988) for Memphis is another application of advanced electrochemical materials for cultural rather than functional innovation (figure 1.4).

But generally, designers have not exploited the aesthetic dimension of new materials with the same energy that engineers have exploited their functional possibilities (to backlight LCD screens in laptop computers reducing their bulk and weight, e.g., or to illuminate escape routes in aircraft so they can be seen through smoke).

Most work in this area does not encourage poetic and cultural possibilities to converge with practical and technical ones. The outcome is a stream of unimaginative proposals. For example, AT&T has applied for a patent for a coating of colored polymer sandwiched between two thin layers of indium tin oxide that changes color when a low voltage is fed through it; the company plans to use it to enable phones to change color instead of ringing.

Although combinations of matter and information might eventually lead to interactive surfaces, giving rise to new channels of communication between people and an increasingly intelligent artificial environment of objects, most smart materials are still under development, are expensive, and use large amounts of energy to operate. The most interesting materials are not available for design

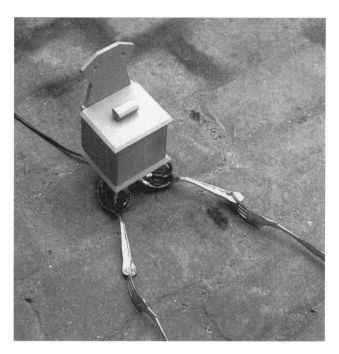

Figure 1.3 Daniel Weil's *Clock* (1983), based on the largest circuit boards available in the early 1980s, is a reaction against miniaturization.

experiments, and one must either use simulations or work with widely available but less sophisticated materials to create emblems of what might be.

Dematerialization

The electronic object is an object on the threshold of materiality. Although "dematerialization" has become a common expression in relation to electronic technology, it is difficult to define in relation to the tangle of logic, matter, and electrons that is the electronic object.

The CPU of an electronic object is, essentially, physically embodied symbolic logic or mathematics. Its "material" representation is the circuit and the components it connects. Symbolic logic describes the workings of the "machine" the object becomes when the program runs. The algorithm is the logical idea behind the program, a strategy that allows symbolic logic to be translated into a programming language (such as C++) and run through the machine, controlling the flows of electrons through its circuitry.

Figure 1.4 Andrea Branzi's *Leaf Electroluminescent Lamp* (1988) for Memphis is a rare example of an application of advanced electrochemical materials for cultural rather than functional
innovation.

Dematerialization, therefore, means different things depending on what it is defined in relation to: immaterial/material, invisible/visible, energy/matter, software/hardware, virtual/real. But the physical can never be completely dismissed: "Every symphony has its compact disc; every audio experience its loudspeaker; every visual image its camera and video disc. Behind every outward image or symbol lies mechanical support, and if the immateriality of these images and symbols gives rise to a new approach to the relationship between human being and object, the analysis will be one of the individual's connection with the material support underlying the new culture of immateriality" (Moles 1995, 274).

One argument, put forward in the 1980s by the design group Kunstflug, is that values and functions can completely shift from hardware to software, from three to two dimensions, and ultimately to "design without an object." It sounds like an untenable and an oversimple critique of materialism, but during the mid-1980s it drew attention to their ideas. They argued for a change in the attitude to the consumption of objects, calling on industry to produce solutions, not commodities. "Design without an object" could, as part of a cultural movement, offer an alternative to abstinence from consumption while encouraging "the forsaking of things as objects of desire and covetousness."

In the exhibition Design Today, held at the German Museum of Architecture in 1988, Kunstflug offered two examples of this approach: design proposals for *The Electronic Room: Programmable Appearances—Surfaces, Appliances, Comfort* (figure 1.5) and *Electronic Hand Calculator* (figure 1.6). While the room seems only to reinforce stereotypical approaches to the impact of electronics on architectural spaces, the electronic hand calculator became an icon for "design without an object," defining one extreme position in the debate about the impact of electronic technology on objects.

This interest in dematerializing the object for social and political reasons is echoed by the "info-eco" ideas of Manzini, Susani, and Thackara who argue that, by focusing on experiences rather than objects, electronic technology can provide services currently offered through discrete products. In the Info-eco Workshops held at the Netherlands Design Institute in 1995, participants developed scenarios on themes such as "Beyond Being There." Dematerialization was used to investigate hypothetical situations in limited scenarios and discover how information technology might satisfy needs normally fulfilled materially. For instance, telematic tools were proposed where the quality of experience they offered would reduce the desire to travel—digital information being easier to

Figure 1.5 The design group Kunstflug's *The Electronic Room: Programmable Appearances — Surfaces, Appliances, Comfort* for the Design Today exhibition held at the German Museum of Architecture (1988).

move than matter. (Reports detailing the results of the workshops are available at www.design-inst.nl/.)

In the introduction to the 1994 Ars Electronica festival in Linz, Peter Weibel describes a another form of dematerialization, "intelligent ambience." It arises from shifting emphasis from the "machine" to its "intelligence," and distributing that intelligence throughout an environment:

Machine intelligence will serve to make the environment more efficient and more intelligent so that it will be able to respond more dynamically and interactively to human beings. The realisation of the concepts of computer aided design and virtual reality will thus be followed by the realisation of computer aided environment and intelligent, interactive, real surroundings. The latter will be referred to as intelligent ambience—an environment based on machine intelligence. One could say: from Tron house to the Tron ambience. (Weibel 1994, n.p.)

Weibel's observations fall between two other views of dematerialization. The first, which belongs to the human factors world, has been referred to as "ubiquitous computing" and is the subject of much research. Dematerialization is

Figure 1.6 Kunstflug's *Electronic Hand Calculator* for the Design Today exhibition held at the German Museum of Architecture (1988): an icon for "design without an object."

seen as a way of providing "transparent" interfaces for computers by embedding the technology in familiar objects and environments and introducing a high degree of automation. At the other extreme is Design Primario,[7] where design effort shifts from hardware to software, and controls levels of light, sound, and temperature to provide sensual environmental qualities. But the aesthetic possibilities of this form of dematerialization have been best exploited by architects: Toyo Ito's design for his *Dreams Room* at the Victoria and Albert Museum in London, was partly motivated by a desire to extend this approach to include information (which he referred to as "active air").

Another form of dematerialization is defined by electronic objects' role as interfaces. With these objects the interface is everything. The behavior of video recorders, televisions, telephones, and faxes is more important than their appearance and physical form. Here design centers on the dialogue between people and machines. The object is experienced as an interface, a zone of transaction.

Although most work in this area tends to reduce the object to a "graphical user interface," a screen, designers are beginning to explore the full potential of the "thingness" of the object. The product becomes virtualized and is represented by a set of physical icons and their various permutations. This could lead to more sensual interfaces than screens and offer new aesthetic qualities.

The work of Durrell Bishop offers a vision of what this might mean: existing objects are used as physical icons, material representations of data that refer to both the pragmatic and poetic dimensions of the data being manipulated. The objects and the electronic structure need have nothing in common. For example, in his design for a telephone answering machine, small balls are released each time a message is left. These balls are representations of the pieces of information left in the machine, allowing direct interaction between the owner and the many possibilities an answerphone offers for connecting to telephone and computer systems. If the caller leaves a number, the ball can automatically dial it; if the message is for somebody else, the ball can be placed in his or her personal tray. Although applied very practically, Bishop's thinking engages with the cultural context in which the technology is used. An "aesthetics of use" emerges.

The material culture of non-electronic objects is a useful measure of what the electronic object must achieve to be worthwhile but it is important to avoid merely superimposing the familiar physical world onto a new electronic situation, delaying the possibility of new culture through a desperate desire to make it comprehensible.

Juxtaposition

How can we discover analogue complexity in digital phenomena without abandoning the rich culture of the physical, or superimposing the known and comfortable onto the new and alien? Whereas dematerialization sees the electronic integrated into existing objects, bodies, and buildings, the juxtaposition of material and electronic culture makes no attempt to reconcile the two: it accepts that the relationship is arbitrary, and that each element is developed in relation to its own potential. The physical is as it always has been. The electronic, on the other hand, is regarded only in terms of its new functional and aesthetic possibilities; its supporting hardware plays no significant part.

Fiona Raby's telematic *Balcony* (figure 1.7) demonstrates how the contradictory natures of electronic and material cultures can coexist. The balcony provides access to an open telephone line linking two or three places. Its physical form provides a focal point and support for leaning on, while an ultrasonic

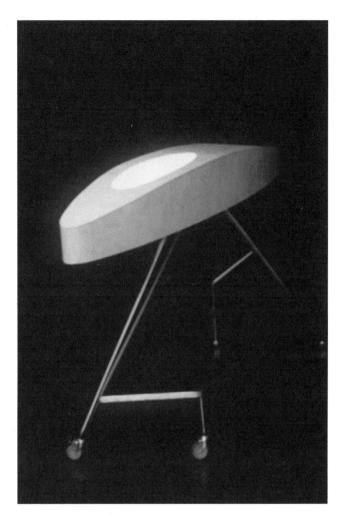

Figure 1.7   Fiona Raby's telematic *Balcony* (1995) is an example of an approach to electronic objects where no effort is made to reconcile the different scales of the electronic and the material.

Figure. 1.8 Robert Rauschenberg's *Oracle* (1965) has had its technology updated three times over thirty years, but its materiality and cultural meaning remain unchanged. Cultural obsolescence need not occur at the same rate as technological obsolescence. Robert Rauschenberg, *Oracle* (1965). Copyright Robert Rauschenberg/VAGA, New York/DACS, London 1999.

sensor detects the approach of users and slowly clears the line. There is no point trying to integrate the physical support and the ultrasonic field, to collapse one into the other, forcing the physical to represent the electronic or to disappear completely so that only electronic effects remain. Juxtaposition allows the best qualities of both to coexist, each with its own aesthetic and functional potential. Technology can be mass-produced whereas the object can be batch-produced. No effort need be made to reconcile the different scales of the electronic and the material. They can simply coexist in one object. They can grow obsolete at different rates as well. Robert Rauschenberg's *Oracle* (figure 1.8) has had its technology updated three times over thirty years, but its materiality and cultural meaning remain unchanged. Cultural obsolescence need not occur at the same rate as technological obsolescence.

Perhaps the "object" can locate the electronic in the social and cultural context of everyday life. It could link the richness of material culture with the new functional and expressive qualities of electronic technology.

In Philips's 1996 *Vision of the Future* project (Philips Corporate Design 1996), a more subtle awareness of the value of material culture has entered the mainstream of design thinking and may well soon enter the marketplace and everyday life. The project consists of over one hundred design proposals for products for five to ten years in the future. But this awareness is primarily expressed in this project by references to existing object typologies—for example, hi-tech medical kits in the form of medicine cabinets—rather than by radically new hybrids. The designers focus more on practical needs, the electronic qualities are not fully exploited, and the types of objects proposed are already familiar from student degree shows. But the designs do achieve a new visual language, sensual, warm, and friendly. They are well-mannered and socially competent. In these projects the electronic object has reached an optimal level of semiotic and functional performance.

The Electronic as Post-optimal Object

The most difficult challenges for designers of electronic objects now lie not in technical and semiotic functionality, where optimal levels of performance are already attainable, but in the realms of metaphysics, poetry, and aesthetics, where little research has been carried out:

This is what differentiates the 1980s from 1890, 1909, and even 1949—the ability of industrial design and manufacturers to deliver goods that cannot be bettered, however much money you possess. The rich find their exclusivity continuously under threat. . . .

Beyond a certain, relatively low price (low compared with other times in history) the rich cannot buy a better camera, home computer, tea kettle, television or video recorder than you or I. What they can do, and what sophisticated retailers do, is add unnecessary "stuff" to the object. You can have your camera gold plated. (Dormer 1990, 124)

The position of this book is that design research should explore a new role for the electronic object, one that facilitates more poetic modes of habitation: a form of social research to integrate aesthetic experience with everyday life through "conceptual products."

In a world where practicality and functionality can be taken for granted, the aesthetics of the post-optimal object could provide new experiences of everyday life, new poetic dimensions.

(In)human Factors

Am I a man or a machine? There is no ambiguity in the traditional relationship between man and machine: the worker is always, in a way, a stranger to the machine he operates, and alienated by it. But at least he retains the precious status of alienated man. The new technologies, with their new machines, new images and interactive screens, do not alienate me. Rather, they form an integrated circuit with me.

—J. BAUDRILLARD, "XEROX AND INFINITY"

In design, the main aim of interactivity has become user-friendliness. Although this ideal is accepted in the workplace as improving productivity and efficiency, its main assumption, that the way to humanize technology is to close the gap between people and machines by designing "transparent" interfaces, is problematic, particularly as this view of interactivity has spread to less utilitarian areas of our lives. According to Virilio (1995): "'Interactive user-friendliness' . . . is just a metaphor for the subtle enslavement of the human being to 'intelligent' machines; a programmed symbiosis of man and computer in which assistance and the much trumpeted 'dialogue between man and the machine' scarcely conceal the premises: . . . the total, unavowed disqualification of the human in favor of the definitive instrumental conditioning of the individual" (135).

This enslavement is not, strictly speaking, to machines, nor to the people who build and own them, but to the conceptual models, values, and systems of thought the machines embody. User-friendliness helps naturalize electronic objects and the values they embody. For example, while electronic objects are being used, their use is constrained by the simple generalized model of a user these objects are designed around: the more time we spend using them, the more time we spend as a caricature. We unwittingly adopt roles created by the human factors specialists of large corporations. For instance, camcorders have many built-in features that encourage generic usage; a warning light flashes whenever there is a risk of "spoiling" a picture, as if to remind the user that he or she is about to become creative and should immediately return to the norm.

By poeticizing the distance between people and electronic objects, sensitive skepticism might be encouraged, rather than unthinking assimilation of the values and conceptual models embedded in electronic objects. I am not arguing for a way of designing that is free from ideological content but, rather, for one that draws attention to the fact that design is always ideological. User-friendliness helps conceal this fact. The values and ideas about life embodied in designed objects are not natural, objective or fixed, but man-made, artificial, and muteable.

This chapter looks at "poeticizing" the distance between people and electronic objects through "estrangement" and "alienation," locating interactivity between transparency and opaqueness, the pet and the alien, prose and poetry. The first section looks at the origins of user-friendliness in human factors and how it manifests itself in design approaches; the second, on transparency, discusses the implications of closing the distance between people and machines; and the third, on (in)human factors, looks at alternatives based on estrangement.

User-friendliness

Manuel DeLanda (1991) situates the origins of the man-machine interface within a military context:

It is at the level of the interface that many of the political questions regarding Artificial Intelligence are posed. For instance, one and the same program may be used to take human beings out of the decision-making loop, or on the contrary, [be] interfaced with them so as to create a synergistic whole. It is the design of the interface which will decide whether the machinic phylum will cross between man and machines, whether humans and computers will enter into a symbiotic relationship, or whether humans will be replaced by machines. Although the centralizing tendencies of the military seem to point to a future when computers will replace humans, the question is by no means settled." (176)

DeLanda writes that research into interactivity between people and computers began with the military's desire to visualize data held in computers, and that interactivity went much further than it intended, giving people total control over their machines. Although scientists such as Doug Engelbert, Alan Kay, J. C. R Licklider, and Murray Turoff managed to gain control of the evolution of computers from the military, developing a vision of interactivity as a partnership between people and machines acted out on the computer screen, they were unable to introduce them into everyday life. It was hackers like Steve Wozniak and Steve Jobs who eventually managed to translate these ideas into a machine that could compete in the marketplace against large corporations like IBM and establish a new model of interactivity.

While interactivity made huge leaps forward before its entry into everyday life through the marketplace, once the computer became a successful mass-produced object, innovation in interactivity shifted from hardware to software, and evolved around screens, keyboards, and mouse-like input devices.

The Human Factors Approach

These days most work on the development of interfaces is by engineers and scientists working for large corporations and universities, and comprising the human factors community. Although mainly concerned with computers, other electronic objects are becoming subject to this approach, particularly as designers have, so far, been unable to develop convincing alternatives.

In a review of *Things That Make Us Smart* by cognitive psychologist Don Norman, Rick Robinson (1994) offers remarks about Norman's view of design that are applicable to the human factors community in general. Robinson argues that Norman's approach results in products that will not confuse or disappoint (which is clearly not enough). His main criticism is that it "misses the essential connection between the power of objects to affect the way in which the world is seen and the mechanism through which that happens. Paradoxically, user-centredness is not just figuring out how people map things, it absolutely requires recognising that the artefacts people interact with have enormous impact on how we think. Affordances, to use Norman's term, are individually, socially, and culturally dynamic. But the artefacts do not merely occupy a slot in that process, they fundamentally shape the dynamic itself" (Robinson 1994, 78).

Design/Aesthetic Manifestations

In the human factors world, objects, it seems, must be understood rather than interpreted. This raises the question: are conventional notions of user-friendliness compatible with aesthetic experience? Perhaps with aesthetics, a different path must be taken: an aesthetic approach might subsume and subvert the idea of user-friendliness and provide an alternative model of interactivity.

The reduction of the relationship between people and technology to a level of cognitive clarity by the human factors community contrasts with the approach taken by Ettore Sottsass in the late 1950s for the design of the ELEA 9003 computer for Olivetti (figure 2.1):

It was immediately obvious in the first years in which I worked on the ELEA that in the design of certain gigantic instruments, as electronic machines were then, or in the design of groups of machines which have a logical and operational relationship between each other, one ends up immediately designing the working environment; that is, one ends up conditioning the man who is working, not only his direct physical relationship with the instrument, but also his very much larger and more penetrating relationship with the whole act of work and the complex mechanisms of physical culture and psychic actions and reactions with the environment in which he works, the conditionings, the liberty, the destruction, exhaustion and death. (Sottsass, qtd. in Sparke 1982, 63)

Figure 2.1 The approach taken by Ettore Sottsass for the design of the ELEA 9003 computer for Olivetti (1959) is very different from the "user-friendly" approach taken by the human factors community, which reduces the relationship between people and technology to a level of cognitive clarity.

Figure 2.2 Marco Zanuso and Richard Sapper's television for Brion Vega was a sophisticated expression of a new role for the skin of an object, with very different characteristics in both its states. Switching it on or off transformed it from familiar to mysterious object.

Although Sottsass's design for a computer clearly derives from a poetic model of people, few designers have developed such powerful aesthetic responses to electronic objects. An exception was Marco Zanuso, whose television for Brion Vega (figure 2.2) was designed with Richard Sapper in 1969 during the high point of the Italian Radical Design movement, and was at the cutting edge of design thinking, a new expression of an everyday electronic product. It took the notion of the black box to the limit, revealing the magic of technology by dissimulating its functional nature. The whole object became a screen, working equally well aesthetically, on or off. Its minimal black form receded when the television image was shown, and it became a pure object when it was switched off. It was concerned with not so much form or even material, but rather the problem of an object with different characteristics in both of its states. It represented a sophisticated expression of a new role for the skin of an object.

Despite this, and because the mechanical design of electronic objects gives few clues to their operation, the problem they posed to most designers soon reduced to one of packaging. But for more experimental designers, the image of the black box became the starting point for exploring new languages of representation rather than interactivity.

Representation

During the early 1980s, in the Department of Industrial Design at the Royal College of Art (RCA) many innovative projects were produced (figure 2.3) that exploited the new freedom offered by the fluid qualities of electronic technologies, although most were still concerned more with representation and interpretation than function or interactivity. As a group these works are impressively diverse, original, and fresh. They imply no clear manifesto or philosophy, but rather reflect the individual personalities and interests of the designers. They explore how different languages of form map onto electronic technologies by reinterpreting existing products. Many of the presentation models were simplified, intended to communicate ideas about form and representation rather than manufacture and practicality. The most relevant work from this era, by Weil, is discussed later in this chapter.

Product Semantics

During the 1980s "product semantics" began to influence thinking about electronic products. Semantics and semiotics were originally used by linguists to understand the structure of language and how it conveys meaning, and later by film theorists (often combined with psychoanalysis—e.g, Laura Mulvey) to analyze how codes and conventions work. In design they were used to analyze the way form could be used to express implicit meanings: the flow of air in a fan heater, for instance (figure 2.4).

Cranbrook Academy's industrial design course developed this approach, led by Michael and Katherine McCoy. From the mid-1980s on, its students fed the international design press a steady stream of conceptual designs for electronic products. In 1987 one of them, Lisa Krohn (with Tucker Viemeister), won a competition to promote and exploit the versatile properties of plastics with her design for an answerphone (figure 2.5). The versatility of plastics in this instance

Figure 2.3 Matthew Archer's miniature computer is one of many projects produced in the Industrial Design department at the Royal College of Art during the 1980s that exploited the new freedom offered to design by the fluid qualities of electronic technology.

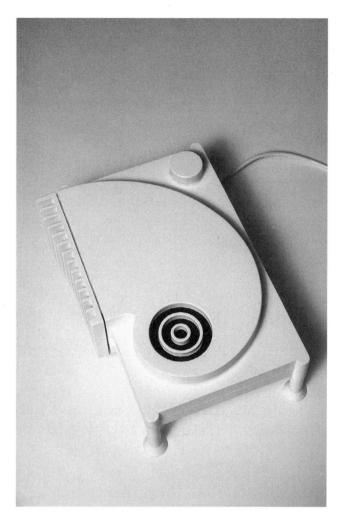

Figure 2.4 Semiotics and semantics were used by 1980s designers as a framework for analyzing the way industrial designers could use form to express implicit meanings: for instance, the flow of air in this fan heater (1981) by Winfried Scheuer.

Figure 2.5   Lisa Krohn's design (with Tucker Viemeister) for an answerphone (1987) shows how a literal use of analogy results in metaphors with a single meaning. Products become depictive of what they do, limiting the viewer's interpretation of the electronic object to the designer's.

is in the area of linguistic expression: "A combination of telephone and answering machine which transcribes and thermally prints any messages, its modern streamlined appearance uses a book format with the pages serving as switches for the different functions" (Form Finlandia, 1987, n.p.). Such literal use of analogy results in metaphors with a single meaning. Products depict what they do, limiting the viewer's interpretation of the electronic object to the designer's, and, although sometimes the link made between groups of objects is ingenious, the power of these borrowed images to sustain interest is weak—they are the material equivalent of one-liners. Once the viewer grasps the connection, there is little else to engage with.[1]

The new forms are just as vigorously tied to their signifieds as the old ones, albeit signifieds extrinsic to the object, based in a cultural frame of reference. To use preexisting patterns of knowledge to define a new technology's possibilities for conveying meaning is not far removed from the Victorian use of Corinthian columns to support beam engines; design holds back the potential of electronics to provide new aesthetic meanings: "Official culture still strives to force the new media to do what the old media did. But the horseless carriage did not do the work of the horse; it abolished the horse and did what the horse could never do" (McLuhan 1970, 133).

Transparency

Because the mimetic approach has greatly affected mainstream thinking about electronic objects, most designs for interfaces with electronic products draw on familiar images and clichés rather than stretching design language. Nothing is what it appears, but simply an allusion to something we are already familiar with. Designers using existing codes and conventions to make new products more familiar often unconsciously reproduce aspects of the ideology encoded in their borrowed motifs. The easy communication and transparency striven for by champions of user-friendliness simply make our seduction by machines more comfortable.

Biomorphism

The trend for forms of biomorphic expression, particularly in cameras and other portable devices, can be seen as expressing either an uncritical desire to absorb technologies into the body, a wish to be a cyborg, or, more optimistically, a need to mold technology to the body. But this need for symbiosis does not have to be expressed through the clichéd language of bio-form; after all, the symbiosis yearned for is often mental not physical. An engaging, if conservative, image of this desire for symbiosis between people and the environment of electronic artifacts is provided by the series of kitchen tools designed by Marco Susani and Mario Trimarchi for the 1992 Milan Triennale. A mixture of abstract form and familiar materials, they neither pretend to have always been there nor are they completely alien (figure 2.6).

For extreme expressions of this wish for transparency or symbiosis, we need to look outside the design field, at the work of the artist Stelarc. He describes a synthetic skin that, absorbing oxygen through its pores and efficiently converting light into chemical nutrients, would make our internal organs redundant and allow them to be removed to create room for more useful technological components. In a performance at the Doors of Perception 3 conference in Amsterdam in 1995, remote viewers were able to manipulate his body into positions

Figure 2.6 Marco Susani and Mario Trimarchi's *New Tools (for the kitchen)* for the 1992 Milan Triennale demonstrates that the need for symbiosis does not have to be expressed through the clichéd language of bio-form; after all, the symbiosis yearned for is often mental not physical.

that represented letters; a computer program allowed sequences to be made up forcing the artist, through electrical stimulation of his muscles, to enact a bizarre semaphore. In an earlier piece, *Third Hand,* he wrote single words with a third artificial hand strapped to one of his own (figure 2.7), activated by the EMG signals of the abdominal and leg muscles, while his real arm was remote-controlled and jerked into action by two muscle stimulators. Stelarc's work illustrates one vision of cyborgs. His work explores the interplay between self-control of the body and its control by the technological logic embodied in prosthetic devices.

Pets

If the desire for familiarity is applied to more complex machines with a potential for autonomous behavior, we could find ourselves living in a bestiary of technological "pets," or zoomorphic electronic objects. Although there is plenty of potential for new aesthetic experiences through the expression of electronic objects' behavior, this area is already dominated by an oversimple mimicry of human and animal behavior. The aesthetic experience they give rise to is based on recognition rather than perception.[2] The users experience something familiar rather

Figure 2.7 In *Third Hand,* Stelarc wrote single words with a third artificial hand strapped to one of his own, activated by the EMG signals of the abdominal and leg muscles, while the real arm was remote-controlled and jerked into action by two muscle stimulators.

than new, so they are conditioned to accept things as they are. Rather than being stimulated to modify their ideas about reality, the users become part of a behavioral "circuit":

The famous Japanese car that talks to you, that "spontaneously" informs you of its general state and even of your general state, possibly refusing to function if you are not functioning well, the car as deliberating consultant and partner in the general negotiation of a lifestyle, something—or someone: at this point there is no longer any difference—with which you are connected. The fundamental issue becomes the communication with the car itself, a perpetual test of the subject's presence with his own objects, an uninterrupted interface. (Baudrillard 1983, 127)

Not all work in this area closely mimics human and animal behavior. *Satori TV* (figure 2.8), a small television that turns it head to face the viewer when touched, is one of the few objects designed at Cranbrook during the 1980s that goes beyond visual semiotics by using performance. This television suggests a life where our only company will be electronic domestic appliances, which must supply the missing banalities of everyday human contact. The artist Alan Rath

Figure 2.8 Peter Stathis's *Satori TV* (1988), which turns its head to face the viewer when touched, suggests a life where our only company will be the electronic appliances of the home, which must supply the missing banalities of everyday human contact.

goes one step further and literally gives technology a face, but not comfortingly. His faces are juxtaposed and recombined with other body and machine parts to create strange and sinister hybrids of people and machines. He uses videos of parts of the face, or whole faces held captive within cathode ray tubes: in *C-Clamp* a face grimaces while its CRT container is held in a C-clamp (figure 2.9). Many of his pieces rely on puns, are comic and anthropomorphic, and remind us of our fear that machines might have lives of their own. But although such works remind us of a possible future where the human soul becomes literally trapped within the machine, their easy appeal means they are also easily forgotten; they are not disturbing enough to shock.

Aliens

A range of possibility exists between ideas of the "pet" and the "alien." While the pet offers familiarity, affection, submission, and intimacy, the alien is the pet's opposite, misunderstood, and ostracized. The artist Martin Spanjaard

Figure 2.9 Alan Rath's *C-Clamp* (1992) literally gives technology a face, but not in a comforting way. His faces are juxtaposed and recombined with other body and machine parts to create strange and sinister hybrids of people and machines.

explores this space, believing: "In order to get used to talking to a machine, one should have one as a pet. A machine which has no particular function, and cannot actually be operated, but which responds to the events in its environment by producing spoken language. Like a cat, which rubs its head against you and meows when it wants to eat or go outside, or a dog which whines when you kick it" (Van Weelden 1992, 247–250).

Spanjaard's robot Adelbrecht evolved over ten years, starting in 1982, from his desire to build a ball that would roll of its own accord and, when it collided with other objects, reverse, change direction, or take other appropriate action. As technology developed so did Adelbrecht; he can now sense whether he is being picked up or stroked, and whether and by how much light and sound are present, influencing his mood or "lust" as it is termed by the artist. Adelbrecht expresses the level of his "lust" by rolling about and by a voice provided by the Institute for Research on Perception in Eindhoven. For example, if he has not been touched since becoming active, on becoming stuck he will call for help; but if he has been touched, he will call his owner. He says "Nice" on being stroked, and "Is it you?" on being picked up. The artist does not program Adelbrecht to totally replicate human or animal psychology, which results in unexpected and quite poetic mumblings. Adelbrecht is an example, as boundaries blur between ourselves and our digital environment, of where a new sense of "alienation" or distance may be discovered. The electronic object does not have to fulfill our expectations; it can surprise and provoke. But, to fulfill this potential, designers need to leave behind a desire to model the new world of electronic products in their own, human, image.

(In)human Factors

If user-friendliness characterizes the relationship between the user and the optimal object, user-unfriendliness then, a form of gentle provocation, could characterize the post-optimal object. The emphasis shifts from optimizing the fit between people and electronic objects through transparent communication, to providing aesthetic experiences through the electronic objects themselves.

But if aliens and user-unfriendliness are to be the alternatives to pets and user-friendliness, this user-unfriendliness does not have to mean user-hostility. Constructive user-unfriendliness already exists in poetry:

The poetic function of language has as its effect that when we read literature we become more aware of language than we are when we are confronted by language in its other functions. To introduce another term dear to the formalists, in literature language is "foregrounded." This, as Jakobson stresses, is the tendency of literature, much more fully recognised in poetry than it is in prose. In the everyday use of language it will seldom be practical and may even be found impolite to "foreground" language. Everyday language is usually informative and instrumental; there is no call for either the speaker/writer or hearer/reader to dwell on the form of what is said/written since if a piece of information has been successfully passed or some action successfully instigated, the words by which this has been managed can count as "transparent." With the poetic function comes a certain opacity, for the writer is no longer passing information nor seeking to instigate action. There may also come an intentional ambiguity." (Sturrock 1986, 109–110)

Defamiliarization

The poetic can offer more than simply enriched involvement. It can provide a complex experience, critical and subversive. The Russian formalist poets of the 1920s based their ideology on estrangement. According to Viktor Shklovsky, the movement's best-known exponent, the function of poetic art is to counteract the familiarization encouraged by routine modes of perception. We readily cease to "see" the world we live in, and become anaesthetized to its distinctive features.

Lebbeus Woods, an architect who has produced imaginary schemes (e.g., Origins) exploring this quality in building, refers to this strangeness as "objectivity," meaning not an analytical state of mind but simply the appreciation of the objects as themselves, independent of the operations of the mind upon them.

The effect of strangeness, infusing an encounter with the unfamiliar and the unknown, was used by Bertolt Brecht to alienate the audience and make them aware that the institutions and social formulae they inherit are not eternal and natural but historical, man-made, and so capable of change through human action. He termed it the "A-Effect," developing the conditions for informed appreciation rather than unthinking assimilation. And Theodore Adorno wrote that authentic art could only function to resist totalization if it was strange and unfamiliar.

Design as Text

Despite an interest in linguistics and texts, the Cranbrook work stopped short of realizing the full potential of the model of meaning it pursued. Rather than radical provocations, it produced beautiful, affirmative designs that were in literary terms structuralist rather than post-structuralist.

Daniel Weil's work, on the other hand, shows what can be achieved if the notion of object as text is taken to its (apparently illogical) logical conclusion, echoing the "death of the author" in literature. His designs challenge the observer to participate in constructing their meaning, with their questions, interpretations, and criticisms becoming part of its meaning.

Weil's designs could be defined as a "text" in Roland Barthes' definition: a "space" of chains and layers of meaning between the object and the viewer, continuously expanding with no fixed origin or closure. When the boundaries of the work are demolished, the text opens out onto other texts. Barthes redefined "text" as a meta-linguistic mechanism that reorganizes the linguistic order, affecting the relationship between writing and reading. Writing and reading, the pre- and post-textual, are of equal value, and both writer and reader are required to exert an equal effort of imagination. Similarly, in the case of a design object as text, designer, and viewer play equal roles. This approach lends itself easily to electronic products, because their components can be freely arranged, unlike mechanical products where the arrangement of components is determined by technical constraints: "In Weil's view the object has a conceptual story which the person owning it has to complete . . . his approach is heavily influenced by Duchamp's conception of the 'unfinished picture'. . . for computer designers, as for Duchamp, the focus of their work now is the process of use of computer systems . . . security is not the objective. He offers a degree of understanding of technology, but control and domination over it are not assured" (Thackara 1996a, 72).

Weil's radios and clocks of the early 1980s are a good example of a research project exploring the aesthetic nature of electronic objects. Most products from this phase of his work seem transient and cheap. Thackara suggests this is an essential part of their nature, as their frailty reminds us of the delicate nature of our conceptual models too. They are objects about objects in the age of electronics, and they express our changing relationship to objects brought about by electronic technologies. They sometimes do this clearly, as in *Four Boxes and One Radio* (figure 2.10), a literal expression of the fact that all radios are packages in a box: the materials have little intrinsic value but acquire value through the authorship of the designer. At other times they do so more obscurely, as in *Small Door,* another design for a radio (figure 2.11). Weil's designs are conceptual and open-ended, and they challenge the user or viewer to engage with them. In literary terms they are post-structuralist.

Like most experimental designs for electronic objects during the 1980s, though, Weil's designs are reinterpretations of existing objects, primarily radios. Perhaps the radio is the electronic equivalent of the chair: a familiar and culturally rich object used by architects and designers as a vehicle to communicate new ideas. Although clocks and radios might seem trivial as technological objects, this is often the only level at which experimental electronic objects can be batch-produced without large investment. Ultimately, the radicalness of Weil's objects lies in their novel imagery and his open-ended approach to meaning. But they still package technology as a visual sign.

Bypassing the Self

Whereas the apparent strangeness of Weil's objects relates to linguistics and notions of the object as text, the architect Kei'ichi Irie and the computer artist Masaki Fujihata use technology to give strangeness to non-technological objects. They explore ways of incorporating technology into processes that bypass our desire to model reality in our own image. The resulting artifacts are sophisticated and subtle fusions of what is and what might be. They map the interface between the social consciousness of the individual designer and the collective scientific consciousness, the dominant ideology embodied within the operating systems of the computer.

Figure 2.10   Daniel Weil's *Four Boxes and One Radio* (1983) is a literal expression of the fact that materials used in the design of cases for radios have little intrinsic value, but acquire value through the authorship of the designer.

Figure 2.11 Daniel Weil's *Small Door* (1986) is more obscure. It challenges the viewer to participate in constructing its meaning. The viewer's questions, interpretations, and criticisms are part of the object's meaning.

As a designer operating in a media-saturated cultural sphere, Irie utilizes computer errors to escape making uncritical and unconscious use of existing cultural forms and conventions, and reproducing the ideology they encode. He considers designing to be autogenerative, made up of subroutines. For Irie, when anything is possible, design is no longer about necessity but becomes a play between subroutines, exploring what can be used rather than realizing an optimum fit. A valid decision may be made on a whim for, as with Weil, the experience of the work is partly what the viewer brings to it: "Even in my own house at Sangubashi, the meaning came from the programming. Which is to say, the elements and methods I employed may have dictated a 70s Tokyo house, but that filter aside, you can see it was just a program. The final form did not have to come out like that at all. If I had applied another filter—who knows?— a tile roof might have resulted" (Irie 1988, 8–9).

Irie's project for a chair (figure 2.12) experiments "with the interplay of noise and unadulterated parts." He first designed a computer program that generated different configurations for a chair with three legs and a seat. The structure of a practical chair is a main routine, but the program generates a host of variants,

Figure 2.12 Kei'ichi Irie's *Lascaux Chair* (1988) began as a design for a computer program. The structure of a practical chair is a main routine; the program generates variants, splitting legs in two, twisting and stretching elements.

splitting legs into two, twisting and stretching elements. The designer simply edits, making selections and adjusting them to ensure they function as free-standing chairs. To Irie's delight, the addition of a number or two to the program can radically change the structure. He uses the computer as an extension of his consciousness: "My thought processes externalised in the form of a chair, which are in turn output as a terminal device 'chair.'"

Irie applied this thinking to his work as an industrial designer with a large housing manufacturer. In his view each company has a "guiding will" program or main routine. When one understands this progam, it is possible to write "bugs" into it, generating objects that are neither the familiar output of large corporations nor the singular expression of the designer as author, but a new, technologically mediated collaboration between designer as virus and industry as program.

Fujihata (1991) responds to Tokyo's unique mixture of immaterial and material culture through an unconventional and conceptual form of industrial design. *Forbidden Fruits* realizes computer visions (figure 2.13), using a CAD system designed for industrial designers and linked to a model-making system.

Figure 2.13 In *Forbidden Fruits* (1991), Masaki Fujihata regards these computer graphic images as "virtual fruit he is forbidden to hold."

An ultraviolet beam traces forms in a photosensitive resin that solidifies on contact with the light, creating translucent representations of computer data. His introduction claims that photography has generated a special "mental software" that is exploited by computer graphics. Interested in going beyond this to discover new potentials for computer graphics, Fujihata transports forms from the screen into the here and now, using a process very different from classical modes of making pictures and sculptures. He articulates data to edit form, using a tree structure to represent the process. On a whim, he returns to points, suddenly turning, constantly producing the tree map of his explorations from which grows "the virtual fruit he is forbidden to hold."[3]

Functional Estrangement

The objects Irie and Fujihata produce focus attention on the design process. They do not challenge the way we experience reality. To provide conditions where users can be provoked to reflect on their everyday experience of electronic objects, it is necessary to go beyond forms of estrangement grounded in the visual and instead explore the aesthetics of use grounded in functionality, turning to a form of strangeness that lends the object a purposefulness. This engages the viewer or user very differently than the relatively arbitrary results of Irie or Fujihata, the crude interpretations and explanations offered through the well-mannered and facile metaphors of mainstream design, or the soft cybernetics of the human factors community. This strangeness is found in the category of "gadget" that includes antique scientific instruments and philosophical toys, objects that self-consciously embody theories and ideas.

The fit between ideas and things, particularly where an abstract idea dominates practicality, allows design to be a form of discourse, resulting in poetic inventions that, by challenging laws (physical, social, or political) rather than affirming them, take on a critical function. Such electronic objects would be conceptual tools operating through a language of functionality that is entangled in a web of cultural and social systems that go beyond appearance.

Although transparency might improve efficiency and performance, it limits the potential richness of our engagement with the emerging electronic environment and encourages unthinking assimilation of the ideologies embedded in electronic objects. Instead, the distance between ourselves and the environment of electronic objects might be "poeticized" to encourage skeptical sensitivity to the values and ideas this environment embodies. This could be done in a number of ways, of which the most promising is a form of functional estrangement: "para-functionality." This quality, common to certain types of gadget, is the subject of the next chapter, which reviews projects and objects that work in this way and explores how para-functionality could be applied to electronic objects.

Para-functionality: The Aesthetics of Use

This chapter reviews projects from art, architecture, and design that exemplify the functional estrangement I call "para-functionality." The term means here a form of design where function is used to encourage reflection on how electronic products condition our behavior. The prefix "para-" suggests that such design is within the realms of utility but attempts to go beyond conventional definitions of functionalism to include the poetic.

Eccentric Objects: Para-functionality and Non-design

Some naive, curious, or eccentric objects, outside the world of conventional design, unintentionally embody provocative or poetic qualities that most product designs, even those intended to provoke, seldom achieve. Although industrial designers play a part in designing instruments of death (weapons) and pleasure (sex aids) these extreme areas of material culture rarely enter design discourse. Yet Jack Kevorkian's *Suicide Machine,* a powerful "unofficial" design that materializes complex issues of law, ethics, and self-determination, shows how an industrial invention can be a form of criticism (figure 3.1). Critical of a legal system that outlaws euthanasia, Kevorkian has his machine to overcome this. Its ambiguous status between prototype and product makes it more disturbing than pure artworks by blurring boundaries between the everydayness of industrial production and the fictional world of ideas. It suggests a role for design objects as discourse where functionality can be used to criticize the limits that products impose on our actions.

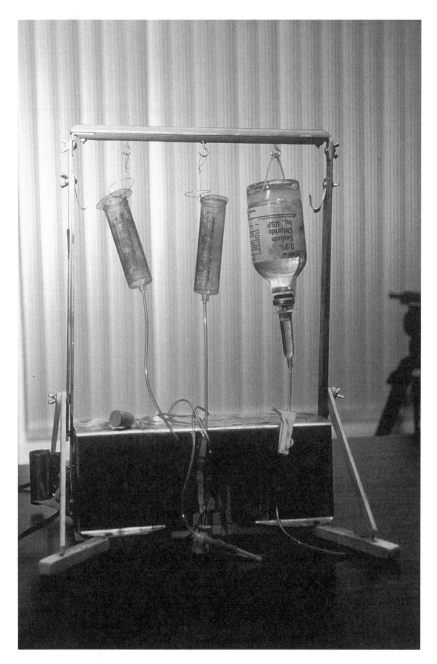

Figure 3.1 Jack Kevorkian's *Suicide Machine* is a powerful piece of "unofficial" design and shows how an industrial object can embody complex ideas through invention as a form of social criticism.

At the other extreme is the world of antique walking sticks. A drinking cane, designed for an alcohol merchant who must spend much of his time visiting the bars of his customers, discretely siphons off his drink while his host is not looking; a trigger later releases the drink into a gutter (figure 3.2). It satisfies etiquette and exploits the walking stick's inherent potential for connection to other objects and contexts: hand, bar, glass, and gutter.

Walking sticks that become a card table or seat (figures 3.3–3.4) show how simple portable props can transform architectural spaces. They conceptually colonize the functional possibilities of preexisting spaces. The user becomes a protagonist in a new narrative where a lobby or park becomes a casino.[1]

A third device, used by detectives in the 1940s for protecting fingerprints on a steering wheel, is beautifully absurd and surreal (figure 3.5). Sigmund Freud (1996, 13) cites G. Heymans's explanation that a joke works through bewilderment succeeded by illumination. The word that is the vehicle of a joke often appears at first to be wrongly constructed, unintelligible, incomprehensible, or puzzling. In this double steering wheel a similar unintelligibility is evident: its comic effect is produced by solving this bewilderment by understanding its function. This is also the case with "Chindogu" (figure 3.6). Their individual elements are recognizable, but the reason for combining them is at first bewildering. The meaning behind the object is derived from "sense-fiction": the objects make functional sense, but are still useless.[2]

Forbidden Emotions: Para-functionality and Design

In a review of an exhibition of work by Intermediate Unit 3, Objects in the Landscape, at London's Architectural Association, Irie (1993) contrasts the "electronic devices essential to contemporary urban existence," the means whereby "information, entertainment and fantasy are promoted—and controlled," with the unit's "virus-like prototypes" that "invade and disrupt such networks, and propel minds and bodies into a hectically deregulated world of fragments—fragments of ideals, of illusions, of sensory impressions." The use of strange inventions by architects is not uncommon and, although they have lost much of their potency through overuse, their deployment in this instance as "bizarre monsters," designed to challenge the banal reality supported by consumer durables, emphasizes the need to identify how electronic products can offer alternative expressions of their own functional logic. In a field where "product design is thoroughly integrated in capitalist production, [and] bereft

Figure 3.2 This drinking cane from the Saint-Etienne mail-order catalogue of 1910 operates in a context where etiquette assumes such importance that the object must be made to maintain it in a "socially dangerous" situation.

Figures 3.3–3.4 The table cane, patented in England in 1891, and the "low seat cane" are examples of how simple portable props can transform an architectural space.

of an independent critical tradition on which to base an alternative,"[3] only a few designers use the function of products as criticism.

For example, Penny Sparke (1982) cites Gaetano Pesce: his "use of distortion and exaggeration [are] 'absurd' devices for commenting upon his observations. Rather than turning to alternative media, Pesce uses the language of design to make its own self-commentary" (52), but his objects do not incorporate functionality as a primary component (figure 3.7). When functionality does enter, it is often jokey and closer to the playful one-off multiples created by Fluxus. During the 1980s Denis Santachiara and Philip Garner developed approaches that merit a closer look. Santachiara, who developed a distinctive approach over many years, aims to raise the aesthetic quality of mass-produced everyday objects such

Figure 3.5 A steering wheel, used by detectives during the 1940s to drive recovered vehicles back to the police station without smudging the thief's fingerprints.

as domestic appliances by developing their possibilities of animation. This could be seen as little more than a desire to use technology to give objects a personality by making them more expressive and quirky (figure 3.8). But his concern is with an aesthetics of use which give objects a distinctive identity from the linguistics of construction and manufacture. Santachiara subverts technical knowledge, redirects it towards provocative ends, provides more than enriched interactivity, and raises the complex issues of what Baudrillard has called the "crisis of functionalism."

Baudrillard (1981) argues that the acceptance of functionalism as an arbitrary but dominant rationality gave rise to an irrational counter-discourse that moves between the two poles of kitsch and surrealism:

The surrealist object emerges at the same epoch as the functional object, as its derision and transgression. Although they are overtly dys- or para-functional, these phantasmic objects nevertheless presuppose—albeit in a contradictory sense—the advent of functionality as the universal moral law of the object, and the advent of this object itself, separated, autonomous and dedicated to the transparency of its function. When one ponders

Figure 3.6 The individual elements of a "chindogu" are recognizable — in this case, a clothes dryer and golf club, but the reason for their combination is at first bewildering.

it, there is something unreal and almost surreal in the fact of reducing an object to its function: and it suffices to push this principle of functionality to the limit to make its absurdity emerge. This is evident in the case of the toaster, iron or "undiscoverable objects" of Carelman. (192–193)

Santachiara's work is often closer to kitsch than that of Garner, whose is closer to surrealism and the absurd. Garner's proposals for products are a form of industrial design that taps into the strange psychological and social narratives arising from the objects themselves and interaction with and through them in a consumer-oriented society. Although their overtly satirical and whimsical

Figure 3.7 Gaetano Pesce's furniture for Cassina during the early 1960s uses the language of design to communicate his observation that people will always be alienated from objects as long as consumption is the primary reason for an object's existence.

character, often simply visual puns or jokes, undermines the viewer's suspension of disbelief (figure 3.9), they demonstrate the power of mock-ups, scenarios, and fictitious narrative over working prototypes as a way of presenting this kind of fiction. The success of both his books confirms that people understand and relate to the narrative behind the work without having to use the objects.

Santachiara and Garner operate within the realm of the gadget, the opposite of the well-designed object. The term "gadget" here denotes a curious, original and witty accessory of no real use, as opposed to the "gimmick," which is too transparent in its effort to impress and attract attention. Giulio Ceppi remarks that "probably the gadget has never been considered, by official design culture, as the result of a design effort, an industrial product capable of revealing interesting technical features or of influencing peoples behaviour" and that "the most important phenomenon caused by the gadget is, however, a psycho-behavioural factor: wonder. . . . The fact that wonder and surprise are two variables that rarely enter into the design of industrial objects has induced the development of a clandestine niche in which such forbidden emotions can be found" (Ceppi 1991, 15).

Heterotopian Gadgets: Para-functionality and Art Objects

For examples that explore the aesthetics of this "clandestine niche" of forbidden emotions it is necessary again to move away from industrial design, and begin

Figure 3.8 Denis Santachiara's *Portale* (1989), which sparks when it is passed through, is an example of his concern with an aesthetics of use where invention is used to give objects a distinctive identity that moves away from the linguistics of construction and manufacture.

Figure 3.9 Philip Garner's *Alienature* (1985) demonstrates the power of mock-ups, scenarios, and fictitious narrative over working prototypes as a way of presenting this kind of fiction.

with literature: not the gadget-ridden world of science fiction but a world where writing itself is a gadget in that it celebrates the workings of language. The heterotopia described by Michel Foucault (1970) illustrates what a literary gadget might be like:

Utopias afford consolation: although they have no real locality there is nevertheless a fantastic, untroubled region in which they are able to unfold; they open up cities with vast avenues, superbly planted gardens, countries where life is easy, even though the road to them is chimerical. Heterotopias are disturbing, probably because they destroy "syntax" in advance, and not only the syntax which causes words and things (next to and also one another) to "hold together." This is why utopias permit fables and discourse: they run with the very grain of language and are part of the fundamental dimension of the fabulous; heterotopias (such as those found so often in Borges) desiccate speech, stop words in their tracks, contest the very possibility of grammar at its source; they dissolve our myths and sterilise the lyricism of our sentences. (xv–xvii)

David Porush (1985) uses terminology that invites comparison between the poetics of real machines and strange inventions, and literary gadgets: "[Samuel Beckett's] *Lost Ones* is a palpable fiction which, even as its inventor attempts to complete the blueprint, collapses into impossible meaninglessness, self-contradiction, and absurdity. The fallibility of the cylinder machine lies in the fact it is constructed in words; the author's attempt to describe it precisely becomes an exercise in the futility of trying to describe anything using language" (161).

Beckett uses two kinds of language, a precise technical/mathematical one, and a language of "failure, probability and doubt." These two rhetorics are at odds with each other and their weaving together provides the qualities of this text, "an allegorical world of pure fiction" about the "perception of the mute resistance of worldly objects to our vain and inappropriate attempts to attach names to them." Paul Klee seems to have incorporated this sensibility into his drawings: for example, *The Twittering Machine* (figure 3.10), where a strange device hovers in the imaginary space of the drawing, suggests a realm where machines do not simply mirror rationality through nonsensical functions but embody alternative physical laws to ours, like Marcel Duchamp's "Large Glass" and the "Pataphysics" of Alfred Jarry.

What happens when this sensibility moves from the page and canvas to become part of everyday space? The sculptor Panamarenko is interesting in this respect as his machines embody the same ambiguity as the literary and painterly gadgets of Beckett and Klee. Whereas artists like Jean Tinguely have constructed useless machines that comically mirror rationality, Panamarenko's objects rarely work (figure 3.11), provoking the viewer to think about the nature of invention and the desires that motivate it. They are about flight, desire, the limits of knowledge, and the transition from wondering and dreaming to the dull reality of realization. By denying that last step and conventional practice, they hover successfully between the imaginary and the real. His scientific theories on flight also highlight the fictional nature of scientific knowledge and blur the boundaries between words and things.

The inventor-artist Steven Pippin meditates on photography. He coats with photographic chemicals the interior surfaces of everyday objects like washing machines, toilets, and bath tubs, turning them into cameras. His ingenious experiments interweave the host object's original functionality with that of a camera, resulting in objects that occupy a difficult conceptual space outside the usual polarization of functionalism and surrealism. They do produce sense, and we understand them, but it is hard to say what exactly we understand about them.

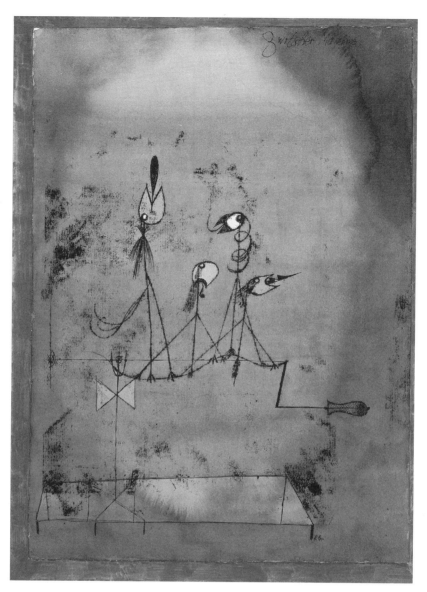

Figure 3.10 Paul Klee's *The Twittering Machine* (1922) shows a strange device hovering in the imaginary space of the drawing, suggesting a realm where machines do not simply mirror rationality through nonsensical functions. Paul Klee, *The Twittering Machine* (1922). Copyright DACS 1999.

Figure 3.11 Panamarenko's *Voyage to the Stars* (1979), like many of his other pieces, does not actually work. This provokes the viewer to think about the nature of invention and the desires that motivate it.

They differ from the symbolic machines and devices of Rebecca Horn, where things do what we expect but the company they keep surprises. Pippin creates conceptual gadgets that render useless our expectation of what things ought to do; they turn knowledge itself into a gadget and allow us to catch glimpses of how knowledge works and wonder at its beautiful but useless mechanisms.

The objects produced by the inventor-artist Philippe Ramette occupy a different part of the space between ideas and things. They resemble in atmosphere the design proposals of Philip Garner but are less ironical in their straightforward presentation of function through the nostalgic language of antique scientific instruments. Meyer Rubinstein (1993) describes them as "prostheses of the spirit" (100), aids to thought and contemplation. As with many of the objects described in this chapter, the emphasis on functionality focuses the viewer's attention on the space between the experience of looking at the work and the prospect of using it. Here the emphasis is on the body and its relationship through the senses to the space that contains it. Although fully working, many of Ramette's objects cannot be used because they can hurt or worse: for example,

Figure 3.12 The emphasis placed on functionality in Philippe Ramette's *Object with Which to See the World in Detail* (1990) focuses the viewer's attention on the space between the experience of looking at the work and the prospect of using it.

Object to Make Yourself Be Struck by Lightning, or *Intolerable Object* whose lens focuses sunlight onto the top of the head. But not all his objects are threatening. In a world of artificial objects shaped almost entirely by functionalism, devices like an *Object with Which to See the World in Detail* do not attempt to escape the dictates of functionalism but instead work from within, extending it to include the poetic and playfully subversive (figure 3.12).

Social Fictions: Para-functionality and Criticism

Although often threatening, Ramette's objects do not shock. Their critical content is hidden beneath the poetry of construction and the humorous appreciation of their function. But the work of the artist Andrea Zittel shocks by using the familiar contexts of the home, and of the system of production and consumption, to concretize alternative values that are outside notions of the future or past but sit uncomfortably alongside "now." They suggest that the way things are may not be the only possibility. They initiate a questioning and awareness that helps unravel the "one-dimensionality" that characterizes present times and maintains "the impossibility of the possible."[4] Zittel's *Comfort Units* suggests an

Figure 3.13 Andrea Zittel's *Comfort Units* (1994) suggests an unusual way of thinking about the role of furniture. Her emphasis shifts from issues of style and image to their psychological use as tools for inhabitation.

unusual way of thinking about the role of furniture (figure 3.13). Her emphasis shifts from style and image to their psychological use as tools for inhabitation. By clearly favoring the manifestation and fusion of particular functional possibilities over others they remind us, through an extreme but credible form of functional reductionism, of our dependence on objects for developing new behaviors. In her work it is never quite clear whether her positive-reinforcement prototypes reflect a genuine belief that this is what we need, or are an ironical play on modernism.

The architects Kenneth Kaplan and Ted Krueger (K/K Research and Development) leave no doubt about the status of their assemblages of found machine parts (figure 3.14) as ironic "analogues" for architectural ideas. Although their writing is polemical their use of objects to attract the attention of the audience,

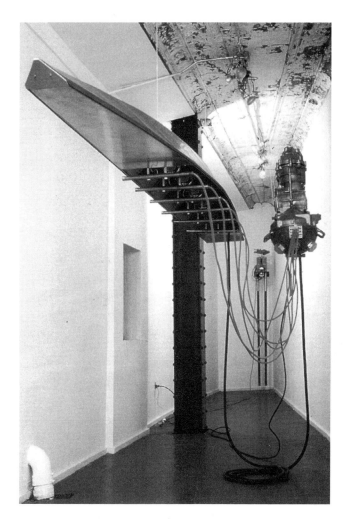

Figure 3.14 K/K Research and Development's *Bureau-dicto* (1989) is an ironic "analogue" for architectural ideas consisting of an assemblage of found machine parts.

Figure 3.15 K/K Research and Development's *Crib-batic* (1986) is a prototype for a push-chair made from steel (they felt children needed to be exposed to hard materials from an early age). It was equipped with measuring equipment so that the child might interact with the environment on the go.

before it is seduced by their usually written political narrative, reduces the objects to dumb props. Their *Crib-batic* project (with Christopher Scholz), however, is an exception (figure 3.15). A prototype for a child's push-chair made from steel (they felt children needed to be exposed to hard materials from an early age), it was equipped with measuring equipment so that the child might interact with the environment on the go. This piece is more powerful than their more obtuse architectural analogues, because it is possible to imagine what it would mean for such thinking to enter everyday life through similar objects. It moves beyond implied functionality and appearances to use function to draw attention to the role objects play in conditioning our responses to the environment.

Another architectural practice, Diller + Scofidio, designs and builds architectural gadgets that work on a critical level. *Para-Site,* an architectural exploration of the impact of electronic media on architectural space, is relevant here because of the equal importance it gives to electronic and conventional media. Electronic objects such as televisions and video cameras are not repackaged or redesigned but integrated into new hybrid objects (figure 3.16), transforming these boring and familiar devices into an architectural intervention. Diller + Scofidio deploy technology intelligently, using it to reveal, enable, and criticize, intervening in not only the abstract space of the building but also its social and practical use.

Para-Site is one of many critical interventions in public spaces by architects and artists. One of the best known is Krzysztof Wodiczko's large-scale projections onto public buildings. He has written: "My socio-aesthetic research and experiences deal with 'strategies' for making public art critical, non-official art." He studied in the graduate program of industrial design at the Akademie Sztuk Pieknych in Warsaw under a former collaborator of Le Corbusier, Jerzy Soltan, who advocated a "(post)-avant-garde" strategy of critical engagement with and infiltration of, the institutional structures of industry and culture. After graduating he worked in Warsaw as an industrial designer for UNITRA, a manufacturer of electronic products. One of his first pieces of art was done in 1969 while still an industrial designer there: *Personal Instrument*[5] (figure 3.17). He was assisted in this by technicians from the Experimental Music Studio in Warsaw:

The instrument transforms the sounds of the environment.
The instrument functions in response to hand movements.
The instrument reacts to sunlight.
The instrument is portable.
The instrument can be used any place and any time.
The instrument is for the exclusive use of the artist who created it.
The instrument permits him to attain virtuosity. (Wodiczko 1992, 76)

Wodiczko has said that "the instrument's magic silence is its socio-political message." Although private, it depends on a public space as a source of sound, and so that others can gaze at it and imagine how it works. According to Wodiczko, "It was a way to shape a metaphor for the limits to the freedom of the individual Pole, to the ways he could exercise this freedom, and to his power in

Figure 3.16 Diller + Scofidio's *Para-Site* (1989), an architectural exploration of the impact of electronic media on architectural space, gives equal importance to electronic and conventional media. Electronic objects such as TVs and video cameras are not repackaged or redesigned, but are integrated into new hybrid objects.

Figure 3.17 Krzysztof Wodiczko's *Personal Instrument* (1969), although private, depends on a public space as a source of sound, and so that others can gaze at it and imagine how it works.

relation to public spaces." It was not designed for mass production nor even for a limited edition "and yet it was intended for the whole world as a metaphor for community life and the nature of public spaces in Poland":

My personal instrument proved to be the point of departure for all my public works. it was my first attempt to provide a metaphorical definition of man's position as a "citizen" of a dominated public space. It was also the first time I attempted to hint at the "strategy" of taking words and using space as medium in which to speak them, even though the right to use a private voice in space that was totally "socialized" (politicized) by the government was utterly nonexistent. I proposed the technique of speaking silently, reticently or by grotesquely exhalting silence." (Wodiczko 1992, 71)

Wodiczko's public projections and homeless vehicles continue this research (figure 3.18). A lesser-known object, *Alien Staff,* shows how industrial design, through conceiving new functions and their configuration as "accessible" products, can function critically. The staff houses a small LCD television, while a small video player, a CB radio or walkie-talkie, and batteries are in a shoulder

Figure 3.18 One of Krzysztof Wodiczko's *Homeless Vehicles* (1988–1989).

bag. The small size of the display, its position at eye level, and its proximity to the alien's face are all important. Once somebody has been attracted, a relationship is perceived between the face within the screen and the actual face of the alien, conceptual barriers are destabilized, and real communication may begin: "It is an instrument that gives the individual immigrant a chance to 'address' directly anyone in the city who may be attracted by the symbolic form of the equipment and the character of the 'broadcast' program" (Wodiczko 1992, 303).

Wodiczko's designs show how simple electronic technologies can challenge preconceptions, but are at the margins of design. Although I see them as design proposals not artworks it seems that, to hold a design view where electronic objects function as criticism, one must move closer to the world of fine art because the design profession finds it difficult to accommodate such research. Objects such as *Personal Instrument* and *Alien Staff,* with their use of simple electronics and their emphasis on invention and social and cultural content, are rare examples of how product design and the electronic object can fuse into critical design.

Hertzian Pathologies: Para-functionality and Electronic Objects

People like to play lotto and people like to use the ATM. Why don't you make it an option in the ATM to say put your money in and say, I'll bet a little bit and see if I can get a little more out, so you ask for twenty dollars, and you push the button, and you could get twenty-five or you could get fifteen.

— JEFF KIPNIS, "ATM COMPETITION"

Another zone of activity outside that of even the exiled designer is "anonymous design," where alternative conceptual models already find expression through electronic artifacts. "Pathological" gadgets are examples of life outside the normal conception of reality; they are design fictions, deviations, and failures and help to maintain the "impossibility of the possible."

Many of these devices concern communication. Most communication technology is oriented toward the individual; it cannot yet support or even encourage more complex social situations. It is point-to-point, one-to-one, not place-to-place. Yet most of this narrow form of communication takes place within that vast field of telematic possibility, the electromagnetic spectrum. The tools and devices limit the possibilities, not the medium. Ironically, many of the more interesting possibilities can be found in "pathological" products based on paranoia and suspicion. Many are designed to "open up" one-to-one channels, transforming private situations into public ones. Scanners, bugs, and detectors illegally "socialize" the world of private telematics. For example, scanners have tuned into wireless baby intercoms enabling "recreational voyeurs" to listen into intimate bedroom conversation.

The radio scanner[6] hovers at the limits of legality (figure 3.19). In the United Kingdom it is legal to make and sell it but, like many pieces of surveillance equipment, not to use it for eavesdropping. It draws attention to what DeLanda has termed the "policing of the spectrum," not a public space but a highly policed and militarized state space. It is one thing to be prosecuted for eavesdropping but, if the information is passed on to a third party and worse, sold, it becomes a serious offense. If sensitive frequencies are found stored in the memory, the owner is likely to be prosecuted. That the radio scanner is a powerful object, entangled with the social and legal systems of society, has been recognized by the artist and musician Robin Rimbaud, alias Scanner: "To Scanner, the world of the personal phone call—an easily tapped medium, especially if you've been building your own radio sets since your teens—represents a far more honest

Figure 3.19   The scanner is an example of a "pathological product" based on suspicion and para-noia designed to open up one-to-one channels, transforming private situations into public ones.

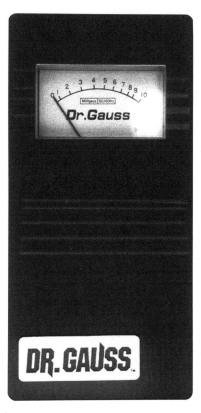

Figure 3.20 The Dr. Gauss EMF detector allows the owner to gather information about the presence of harmful electromagnetic fields so that a complaint can be made.

depiction of the world than the outpourings of televisual reality. And Scanner's records, packed with a huge collection of telephone 'normality,' are, in turn, far more real and disturbing than any arty fabrication of reality" (King, n.d., 136). The radio scanner enables new urban maps to be made, revealing normally hidden structures of the visible and conventional. The scanner is a meta-radio: it transcends the many categories of radio incorporated into commodities, highlighting their commonality as parts of an electromagnetic spectrum.

The Dr. Gauss EMF detector is one step further up the evolutionary ladder of gadgets (figure 3.20), a low-cost version of a usually expensive piece of equipment, used to measure the magnetic component of possibly harmful electromagnetic fields in the home. The device is simply a black box, but the act of using it reveals its conceptual power: when it picks up a field it screams, rising in pitch with the strength of the field.

Objects like this allow us to develop new conceptual models about our environment, helping us to see invisible structures and patterns. They often occupy the cultural wasteland of in-flight magazines, Sunday supplements and specialist shops, where alternative world views embodied as material reality exist as a nonserious and marginal phenomena. But in showrooms they become vital alternatives to art works and galleries. Whereas people step out of ordinary life into an art gallery, the contents of showrooms relate directly to everyday life in the mind of the window shopper.

Between Rationality and Reality

The most effective examples in this chapter function as test pieces that, through their marginalization, make visible the barriers limiting poetic experience in everyday life. The apparent unusability of many of these objects creates a heightened sense of "distance." This can be because the objects do not work technically or, because they are conceptually difficult to assimilate. To see that they are usable is to acknowledge that existing notions of functionality have been extended, a result of imagining uses for these objects. They challenge the impossibility of the possible. It is not enough to look and decode their visual iconography: they must be used. Through use, or at least by modeling a scenario of use in the mind, the observer discovers new ways of conceptualizing reality. They dismantle conceptual models that limit the way we use artifactual reality to extend our scope for action. They challenge how we think about extensions to our "selves" in ways that do not simply magnify but, rather, transform our perception and consciousness of our relation to our environment.

They share no coherent theory. They are simply stories, but stories that allow complex interactions between reality and imagination. Driven by poetry, imagination, and intuition rather than reason and logic, they have their own rationality, an alternative to our everyday scientific-industrial rationality. These are stories about the space between rationality and reality, which in an industrial society have come to be synonymous. When these props are introduced into everyday life as a "virus," subverting it, people can participate in the story, exploring the boundaries between what is and what might be. This is the role of the para-functional as criticism.

By imagining the object in use, we become lost in a space between desire and determinism. Within this space lies the bizarre world of the "infra-ordinary," the subject of the next chapter, which reviews a number of projects in relation to behavior and narrative.

4

Psychosocial Narratives

"Psychosocial narratives" refers to the unique narrative potential of electronic products, the world of desire and fiction that embraces consumer goods, the socialization that the use of electronic products encourages, and the idea that behavior is a narrative experience arising from the interaction between our desire to act through products and the social and behavioral limitations imposed on us through the conceptual models they impose. For instance, although an essential part of everyday life, the telephone embodies crude concepts of social etiquette compared to furniture and architectural space.

This chapter looks at the following ideas: the user as a protagonist and coproducer of narrative experience rather than a passive consumer of a product's meaning; how the psychological dimensions of experiences offered through electronic products can be expanded to include darker conceptual models of need—usually limited to cinema and literature—by referring to the world of product misuse and abuse; the lack of work by authors and filmmakers exploring this area, despite its prevalence in everyday life; the idea that the designer, in their role as a provider of new behavioral opportunities, becomes an "author" working in a medium that can present experiences rather than represent them; and how the electronic product becomes a "role model" bringing about transformations of perception (and conception) in the user as a protagonist by embodying unusual psychological needs and desires in "pathological" electronic objects.

User as Protagonist

The phone and the film projector surely need us to bring them to life, to dial the number or to flick the switch, yet when these machines take us to places, people, and ideas otherwise forbidden, so they flick the switch on us. The beauty of new technology is that by engaging our imaginative co-operation it moves a flat two-dimensional relationship of subject/object, man/machine through a magical door into new mental territory.
—C. BEEVOR, "BETWEEN HERE AND NOW"

The narrative possibilities suggested in this quotation differ from the conventional narrative dimensions created for the electronic product by applying semiotics. Examples of the latter are the Sports Walkman produced by Sony in the mid-1980s, which referred to imagined contexts of use, and the O-Product camera designed by Water Studio for Olympus—in both of which form and texture are manipulated to evoke a world of fantasy and fiction, blurring distinctions between everyday life and the hyperreality of advertising and soap opera.

One way of viewing this difference can be found in *The Meaning of Things,* which outlines an approach to aesthetic experience developed by Rochberg-Halton and based on Dewey's 1934 distinction between perception and recognition:

For Dewey, recognition describes a falling back on some previously formed interpretative schema or stereotype when confronted with an object, whereas perception involves an active receptivity to the object so that its qualities may modify previously formed habits or schemes. Although the explicit purpose of art is to evoke aesthetic experience, Dewey does not limit aesthetic experience to art alone but considers it a potential element of all experience. Perception is essential to aesthetic experience and leads to psychological growth and learning. Recognition, or the interpretation of an object or experience solely on the basis of already existing habits, only serves to condition a person further to a life of convention. If culture were simply a symbol system of convention, as some cognitive anthropologists argue, then aesthetic experience would only consist of recognition in Dewey's sense, because the object of that experience 'contains' meaning only as an arbitrary sign endowed with meaning by cultural convention and not because of unique qualities of its own. (Csikszentmihalyi and Rochberg-Halton 1981, 176–177)

The narrative possibilities offered by the conventional semiotic-based approach depend on "recognition," whereas the more dynamic form of narrative suggested by Beevor could open the way for the active critical receptivity of an experience that "perception" involves.

In the case of electronic products, the "unique qualities" of the object of interaction is their potential as an electronic product to persuade the users as protagonists, through the user's use of the object, to generate a narrative space where the understanding of the experience is changed or enlarged. By using the object, the protagonist enters a space between desire and determinism, a bizarre world of the "infra-ordinary," where strange stories show that truth is indeed stranger than fiction, and that our conventional experience of everyday day life through electronic products is aesthetically impoverished.

The Infra-ordinary

The machine does what the human wants it to do, but by the same token the human puts into execution only what the machine has been programmed to do. The operator is working with virtuality: only apparently is the aim to obtain information or to communicate; the real purpose is to explore all the possibilities of a program, rather as a gambler seeks to exhaust the permutations in a game of chance. Consider the way the camera is used now. Its possibilities are no longer those of a subject who "reflects" the world according to his personal vision; rather, they are the possibilities of the lens, as exploited by the object.
—J. BAUDRILLARD, "XEROX AND INFINITY"

In 1994 the British mobile phone company Cellnet produced a booklet, *Mobile Moments: A Collection of Tales for the '90s,* a chronicle of events that, it felt, demonstrated the crucial part the mobile phone has come to play in our lives. The tales are arranged under headings such as "Mating by Mobile," "Mobile Heroes," "Mobile Marvels" and, most interesting for this chapter, "Mobile Mishaps." Each story is an example of the narrative space entered by using and misusing a simple electronic product, of how interaction with everyday electronic technologies can generate rich narratives that challenge the conformity of everyday life by short-circuiting our emotions and states of mind. I am recommending, not that designers try to predict misuses of products, but rather that they refer to as a context of use this rich narrative space instead of the models of normality usually referred to when new functional possibilities are being developed. As Baudrillard (1990) writes:

In my opinion, the really interesting relations between people don't occur in the form of communication. Something else happens: a form of challenge, seduction, or play which brings more intense things into being. By definition, communication simply brings about a relationship between things already in existence. It doesn't make things appear.

And what is more, it tries to establish an equilibrium—the message and all that. Yet it seems to me that there is a more exciting way of making things appear: not exactly communication, but something more of the order of challenge. I'm not sure that this would invite an aesthetic of communication strictly speaking.

Some people already exploit the potentially subversive possibilities of this parallel world of illicit pleasures stolen from commodified experience. They seek out (ab)user-friendly products that lend themselves to imaginative possibilities for short-circuiting the combinatorial limits suggested by electronic products. In "Fexy Facts," Alfred Birnbaum (1991) writes about the abuse of handheld scanners/fax machines (figure 4.1) by perverts to scan parts of their bodies through sheets of clear plastic and fax the resulting distorted images to lone women. Another example is Douglas Gordon, who appeared on television and used two telephones to call two galleries, fixing the phones together and recording the resulting surreal conversation.

These stories form part of a pathology of material culture that includes aberrations, transgressions and obsessions, the consequences of and motivations for the misuse of objects, and object malfunctions. This is related to the conceptual strategy explored by John Cage in "Radio Music" (1974) and his many pieces for prepared pianos, and by Nam June Paik in his *Magnetic TV* (sculpture) and *Random Access* audiotape wall (installation). Both artists show behaviors toward technology that invite others to follow. Concerned with software not hardware, they invent new uses for existing technologies and promote interaction with "designed" objects that subvert their anticipated uses. In doing so, they challenge the mechanisms that legitimise the conceptual models embodied in the design of the product or system (piano, television, or tape machine). In his video *Making Do and Getting By* (1995), sculptor Richard Wentworth documents our natural ability to subvert object types and act in new ways on our environment. Often, as a by-product of trying to solve a practical problem, a poetic result is achieved, as different ideas, embodied in objects but usually kept apart, come together to reveal hidden similarities. Cartoons and comedies also present a world where the conventional use of everyday objects is turned on its head, leading to surreal and of course humorous situations.

One of the Frankfurt School's arguments was that pleasure has been desublimated and is only available through buying consumer goods. When an object's use is subverted, it is as though the protagonist is cheating the system and deriving more pleasure than is his or her due. The subversion of function relates

Figure 4.1 This handheld scanner/fax is an example of an (ab)user-friendly product used to exploit the potentially subversive possibilities of the parallel world of illicit pleasures stolen from commodified experience.

to a breakdown of order; something else becomes visible, unnameable, unable to find a correspondence in the material world. This subversion of function is related to not being able to find the right word, creating neologisms that bend language to accommodate something new. Desire leads to a subversion of the environment creating an opportunity to reconfigure it to suit our "illegitimate" needs, establishing new and unofficial narratives.

Although the misuse and abuse of everyday objects is related to the anthropological study of material culture there is little literature on its surreal aspects. Occasional overlaps with urban legends[1] establish connections with anthropology proper, and offer at least some analysis of this subject, although emphasis is on collating, and discussing the truthfulness of, the legends themselves. In *Mythologies,* Barthes famously introduced the role played by commodities in the formation and consumption of popular culture, but predates the explosion of electronic products which shape nearly every aspect of modern life.

The almost unbelievable stories reported in tabloid newspapers testify to the unpredictable potential of humans to establish new situations despite the constraints on everyday life imposed through electronic objects. A mother shoots her son after an argument over which television channel to watch; the police set a trap for scanner snoopers by broadcasting a message that an UFO has landed in a local forest (within minutes several cars arrive and their scanners are confiscated); a parent is outraged by a speaking doll, made in China, that appears to swear.

A Pathology of Material Culture

The aesthetic potential of the narrative space centered on the consumer product has received surprisingly little attention from artists and writers and even less from designers. Few films or stories acknowledge how our lives and identities are intertwined with machines and artifacts, particularly everyday electronic products. Though we inhabit an environment of electronic gadgets and gizmos, little effort is turned toward exploring what this means. The film *Family Viewing* (1987) by the director Atom Egoyan studies the relationships between members of a family mediated through everyday technologies such as the telephone and the video recorder. Their sometimes unconventional use of these banal technologies is seamlessly integrated into their lives. This encouraging vision of technology, where new media allow additional forms of expression for everyday desires, offers an alternative to Hollywood's sci-fi and shock/horror visions of technology.

Whereas *Family Viewing* focuses on technologically mediated relationships between people, Nicholson Baker's novels *The Mezzanine* (1986), *Vox* (1992), and *Room Temperature* (1990) richly exemplify how product-centered narratives can evolve from interactions with everyday objects. Deploying a refined appreciation of mass-produced material culture, he weaves playful narratives around psychological and physical interactions with and through the most banal artifacts, laying bare the usually hidden mechanisms of everyday material pleasures. In *Room Temperature,* for example, he writes about how as a child, by repeatedly cycling over an electronic-traffic counting cable, he might have influenced the future traffic system of his city. At the same time he exposes the failings of an over-scientific traffic-planning method based on mathematical models rather than "non-mathematical acts of judgement based simply on years of driving indignantly around."

Dumb Type, a group of artists, writers, and architects based in Kyoto, have developed performances about the cultural and behavioral aspects of consumer technology, and commodification's impact on our lives. *The Performance pH,* staged in a pit below the audience, involves visual projections, sound, and dance. A special piece of stage machinery continuously "scans" the stage floor forcing the performers to either jump or dive to the floor as it passes. The event is a multilayered dense montage, whereas in most techno-art performances technology is used mainly to create spectacle. In *Performance pH,* it felt as though the viewer were exposed directly to the affects of technological consumerism.

Although the work of Scanner is usually discussed in relation to surveillance, another aspect of his work draws attention to the psychological space of the airwaves. Telephone conversations represent for him a more honest depiction of the world than the outpourings of televisual reality. Long samples of telephone "normality" "contain more soap opera in the 54 and a bit minutes of Scanner . . . than there is in a month of Eastenders" (King, n.d., 134–138).

In *Crash* the psychopathological nature of everyday technology is explored through a consumer product, the car. Ballard's provocative introductory essay paints a dark picture of the writer's imagination as an " 'inner space,' that psychological domain (manifest for example, in surrealist painting) where the inner world of the mind and the outer world of reality meet and fuse." In a world "ruled by fictions," the writer's task is to invent the reality (Ballard 1990, 5).

Designer as Author

When we talk about what goes on in a computer, we're talking about an entire complex of relations, assumptions, actions, intentions, design, error, too, as well as the results, and so on. A computer is a device that allows us to put cognitive models into operational form. But cognitive models are fictions, artificial constructs that correspond more or less to what happens in the world.
—G. Chapman, "Making Sense Out of Nonsense"

Conventional roles for design include addressing problems set by industry, designing interfaces that seduce the user into cybernetic communication with the corporate cultural values embodied in the emerging environment of digital objects, and finding novel applications for new technologies. But design could also develop new attitudes to electronic technology. To do this, designers could become more like authors, drawing from the narrative space of electronic object misuse and abuse to create alternative contexts of use and need.

Design could explore the fluid interface between "cognitive models [as] fictions, artificial constructs" and new electronic technologies. Designers could create new critical artifacts that help consumers, as protagonists rather than users, to navigate through the "communications landscape" we share with "the spectres of sinister technologies and the dreams that money can buy" (Ballard 1990, 5).

To explore this new role for designers, it is necessary to turn first to architecture. Although the relation of narrative to space has been thoroughly theorized by architects such as Tschumi, and explored through design proposals by Nigel Coates and *Narrative Architecture Today* (*NATO*),[2] few architects have considered narrative space within the context of an electronic consumer-driven society and even fewer in a way that specifically addresses experiences centered on electronic products. Michael Sorkin (1992) comes close on an urban scale, and Catrina Beevor (1983) evokes consumer electronics as a potentially liberating force for narrative architecture.[3] But only a handful of projects have developed actual architectural proposals, usually through the ploy of designing "a house for . . ."

One such project is an early exploration of the poetics of electronically mediated architectural space, in this case urban space, by Architekturbüro Bolles + Wilson. This competition entry uses architectural function to respond lyrically to electronic media through the design of a retreat from the "electronic glare" of Tokyo, an electronic shadow (figure 4.2): "Encompassed in the concepts of 'electronic glare' and 'electronic shadows' is Ninja Architecture. Devised by Architekturbüro Bolles + Wilson Ninja Architecture describes the function and meaning of projects developed in response to an electronically-dependent consumer society . . . an architecture in search of method and meaning" (Dollens 1991, 123).

Appliance House, a project in the form of a book by Ben Nicholson, is another architectural fiction dealing with imaginary psychological narratives, derived on one level in this case from the products in a Sears catalog and on another from the imagined structures built by a kleptomaniac (figure 4.3). The development of the project through collage means that the reality of a built project is unlikely to match the richness of the book and, when we do see glimpses of realized pieces of furniture, only the figurative elements make the transition from book to object.

Toyo Ito's interest in an architectural response to consumerism sets his work apart from most other architects, including those who address technological issues. He deals with fiction, packaging and the private/public experiences of the city. For example, his proposal for a *Dwelling for Tokyo Nomad Woman* (figure 4.4) is very different from the 1966–1968 visions of Mike Webb's *Cushicle and Suitaloon.* While Webb's nomads carry all their belongings with them, Ito's office girl lives in the city, and her home is merely a floating canister for housing the most basic activities, not belongings. Her life revolves around shopping, and her consumerism generates an architecture that offers a fresh alternative to obsessions with the imagery of consumption. In this architectural fiction,

Figure 4.2 This project by Architekturbüro Bolles + Wilson (1988) is an architectural exploration of the poetics of electronically mediated architectural space, in this case urban space.

Figure 4.3 Ben Nicholson's book *Appliance House* (1990) is an architectural fiction dealing with imaginary psychological narratives derived on one level from the products offered in a Sears catalogue and on another from the imagined structures built by a kleptomaniac.

conveyed through photographs she drinks tea, reads magazines, and applies makeup—hardly utopia. Through projects like these, architects explore the psychological and behavioral dimensions of consumer culture rather than the technical, formal, or structural possibilities of consumer technologies.

Product as Role Model

Examples of how design responds to the psychological and behavioral dimensions of electronics can be found at the edges of anonymous design. Obscure marketing and novel technical possibilities lead unintentionally to objects that, although sometimes gimmicky, offer unusual narrative possibilities. A remote-control watch, addressing an unlikely psychological obsession, speaks of a sad need to control the plethora of domestic gadgetry in not only one's own home but also those of others.

Truth Phone™ (figure 4.5), a relatively straightforward example of an object embodying a pathological model of behavior, paranoid suspicion, combines voice stress analyzer and telephone, allowing the user to tell whether the person at the other end is lying. Although it resembles the absurdist gadgets of Garner,

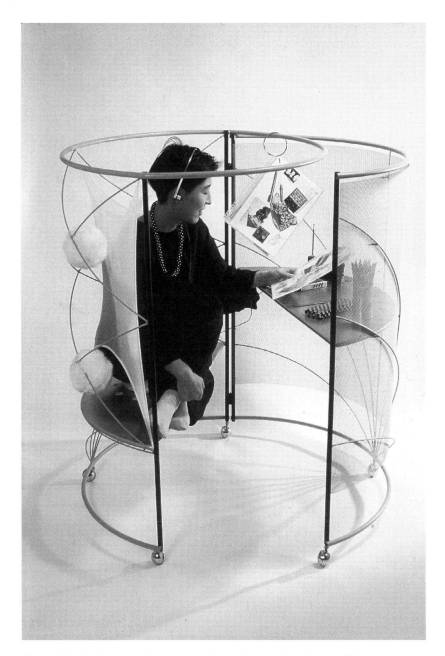

Figure 4.4 Toyo Ito's *Dwelling for Tokyo Nomad Woman* (1985) is an architectural response to consumerism that sets his work apart from many other architects, including those who address technological issues.

Figure 4.5 *Truth Phone*™, by the Counter Spy Shop, is an example of an object embodying a pathological model of behavior. It combines a voice-stress analyzer and telephone, allowing the user to tell whether the person at the other end is lying.

its functional restraint and sober appearance help suspend one's disbelief, something not achieved by many gadgets. *Truth Phone* illustrates how an electronic product can transform the perception (and conception) of the user as a protagonist, in this case by embodying unusual psychological needs and desires in pathological objects. When one imagines using this object to talk to lovers or family members, its critical function becomes clear. By imagining living with it, the owner explores boundaries between himself or herself and the paranoid user-model embodied in the product.

Truth Phone is a "role model" in the sense meant by George Herbert Mead.[4] Through the conceptual model of behavior embodied in its functionality and operation, it allows the user to participate in situations that encourage critical reflection on the socializing effect of our encounters with everyday electronic products. It does this not didactically but in a more ambiguous and indirect way. This and similar electronic objects generate a conceptual space where interactivity can challenge and enlarge the scheme through which we interpret our experiences of using everyday electronic objects and the social experiences they mediate.

Real Fiction

Considered as an operator acting in relation to the daily environment, the designer's ultimate responsibility can only be to contribute to the production of a habitable world, a world in which human beings do not merely survive but also express and expand their cultural and spiritual possibilities. The term habitable, referring to the environment, indicates a complex existential condition that cannot be reduced to its functional component. It is a condition arising from the intersection of a multiplicity of questions rooted in the anthropological and social nature of the human race.

—E. MANZINI, "PROMETHEUS OF THE EVERYDAY: THE ECOLOGY OF THE ARTIFICIAL
    AND THE DESIGNER'S RESPONSIBILITY"

To "contribute to the production of a habitable world," design needs to be transformed, expanding its scope to include speculation on how best to provide the conditions for inhabitation. It must not just visualize a "better world" but arouse in the public the desire for one. Design approaches are needed that focus on the interaction between the portrayed reality of alternative scenarios, which so often appear didactic or utopian, and the everyday reality in which they are encountered.

Many issues touched on here, such as art's relation to everyday life, and the need for art to resist easy assimilation, overlap with those already addressed by the Frankfurt School and others in relation to disciplines such as music (Adorno), painting (Marcuse), art (Benjamin), and drama (Brecht). The similarities between these issues and those addressed by Marxist approaches to aesthetics do not imply an identification with Marxism but are the result of seeing design as having value outside the marketplace—an alternative to fine art.

This kind of design can only exist outside a commercial context and indeed operates as a critique of it. It is a form of "conceptual design"—meaning not the conceptual stage of a design project, but a product intended to challenge preconceptions about how electronic products shape our lives. This chapter discusses how such design thinking might reenter everyday life in ways that maintain the design proposal's critical integrity and effectiveness while facing criticism of escapism, utopianism, or fantasy. The challenge is to blur the boundaries between the real and the fictional, so that the visionary becomes more real and the real is seen as just one limited possibility, a product of ideology maintained through the uncritical design of a surfeit of consumer goods.

The fact that this form of conceptual design need not conform to the conventions that shape the design process in relation to the marketplace does not mean it has to be utopian. It can use its independent position to provide conditions that encourage more reflective and challenging design ideas than are possible in commercial design. But if it is to avoid accusations of escapism this design thinking must also develop strategies for linking itself to everyday life that complement those of the marketplace. This chapter focuses on the problem of "crossing over" and discusses how conceptual design intentions and formats of work, differ from those of commercial design, and require different contexts in which the design thinking can be encountered by the public. It is concerned with representation and contexts of presentation for ideas about everyday life in the form of conceptual design objects.

The Design Object as Prototype

There is a danger that if design is not oriented to the marketplace it is seen as invalid, irrelevant, or self-indulgent, especially if displayed in a gallery. But what if the gallery were viewed as a test-site for designs unlikely to enter everyday life? What would be the most effective format for a design object designed to be shown in a gallery?

The most obvious would be fully working prototypes that can be "tested out" on the public in the gallery and, if the reaction is good, later mass-produced. But fully working prototypes displayed in galleries rarely challenge viewer's assumptions about the role of products in their lives. For example, many visitors to my contribution to the Monitor as Material exhibition at the RCA in 1996 (figure 5.1) said they found the work interesting as spectacle, but had missed concerns with the more fictive, social, and aesthetic aspects that linked it to everyday life, even if only conceptual. Its strangeness and apparent interactivity

Figure 5.1 In my contribution to the Monitor as Material exhibition at the RCA in 1996, many visitors to the space found the work itself interesting as spectacle, but concerns with the more fictive, social, and aesthetic aspects that linked the work to everyday life, even if only through the imaginary, were lost.

emphasized the here and now. The gallery became a "bracketed space," an abstract setting, disconnecting the experience of engaging with the work from everyday life. Displaying a fully working prototype in a gallery context invites people to marvel at the ingenuity of the designer, and the fact it works, but overlook the challenge to the status quo its insertion into everyday life might bring about.[1] Following this route, the gallery becomes a "freak show" of objects of wonder and amusement. The electronic objects of Weil, reinterpretations of existing products such as radios, digital clocks, and calculators, focus on the conceptual relationship between the person and the electronic object. Displayed in the gallery as one-offs, as objets d'art, they achieve little. But if the prototypes are batch-produced (which Weil's objects were), the gallery becomes a "showroom," allowing them to enter everyday life through the marketplace: a specialist shop selling state-of-the art material culture, trading in shock-of-the-new reinterpretations of familiar objects.

The Design Object as Installation

For the designer who regards the electronic object as an embodiment of potential patterns of behavior and ideology, careful consideration of the relationship between the gallery and the conceptual design object is essential if the object is to connect with everyday life.

Electronic objects that use the gallery to demonstrate their interactive aesthetic or experiential aspects can be subsumed by kinetic art culture whose focus is on the here and now and providing an escape from everyday life. An installation by Fiona Raby for Electra 96 highlights this problem. The installation was intended to be a test-site for a design proposal linking two locations by open telephone lines. Ultrasonic sensors registered approaching people and allowed sounds from the other location to filter through, distorted at first they cleared as the person moved closer to the installation so that spoken communication could take place. As a design proposal it would be experienced by a building's inhabitants over several years, and the aesthetic experience would have to be very subtle. As an exhibit in an electronic event the installation was expected to provide immediate feedback in an entertaining way. It might have been better to exhibit a film that used nonworking props to explore how the proposal might impact over time on the day-to-day experiences of fictional users.

One of James Turrell's projects, *Perceptual Cells,* offers an interesting solution to this problem. Once inside a booth-like structure, a bit like a telephone box in a gallery, visitors are presented with controls to vary the color of light in a hemisphere surrounding their heads. Humorous and quirky, it invites comparison with street furniture and public utilities, and their association with mass consumption, state ownership, and industrial production. The visitor imagines, perhaps, using one of these machines on the street, so a strong link with the world outside is established. It successfully combines the best qualities of prototypes and installations: it can be used in the gallery rather than just contemplated, and at the same time establishes links with life beyond the gallery.

The Alien Staff by Wodiczko demonstrates another approach—intervention. Wodiczko's project shows how industrial design, by imagining new functions and configuring them as usable prototypes, can function critically outside the gallery. Wodiczko has deployed teams of "aliens" in various cities armed with his *Alien Staff* and studied the resulting interactions between them and the public (figure 5.2).

Such objects, using simple electronic technologies and emphasizing invention and social and cultural content, are rare examples of how product design and the electronic object can fuse into design as criticism. The prototype draws attention to the boundaries of normal behavior and thought by intervening in everyday social situations outside the gallery. That they are deemed problematic by the design world draws attention to other boundaries of categories of practice and ideas:

Asked how the design world has responded to his various Homeless Vehicle [sic] Wodiczko throws back his head and laughs at the pretensions of the so-called "designer decade." . . . "The minute you present a proposal, people think you must be offering a grand vision for a better future." They can't see a thing like the Homeless Vehicle or the Poliscar as the "concretisation" of a present problem, a makeshift transitional device, or an aesthetic experiment. Instead, "they think it must be designed for mass production, and instantly imagine 100,000 Poliscars taking over the cities." (Wright 1992, 272–273)

The Design Object as Model

What is the potential of nonworking design models as opposed to prototypes? The preoccupation with product semantics, that dominated design in relation to electronic objects for most of the 1980s, focused attention on the object itself, particularly its visual meaning. The concept model functioned as a didactic design object; it was not something to challenge the way we lived our lives, but a meta-design challenging only design itself.

Figure 5.2   Krzysztof Wodickzko's *Alien Staff* (1992) houses a small LCD television. The small size of the display, its position at eye level, and its proximity to the alien's face are all important.

The nonworking model is the conventional physical representation of conceptual design proposals: naturalistic, nonworking mock-ups simulating the appearance of a mass-produced object. Yet this freedom from technical functionality could be better used. If the design model was viewed as a medium in its own right, it could exploit its nonworking status to address issues beyond the scope of the technically functional prototype. But to achieve this it needs to be considered as a model in the same sense as a mathematical or cognitive model. This enlarged view of the model is already accepted in architecture and fine art:

> The space of the model lies on the border between representation and actuality. Like the frame of a painting, it demarcates a limit between the work and what lies beyond. And like the frame, the model is neither wholly inside nor wholly outside, neither pure representation nor transcendent object. It claims a certain autonomous objecthood, yet this condition is always incomplete. The model is always a model of. The desire of the model is to act as a simulacrum of another object, as a surrogate which allows for imaginative occupation. (Hubert 1981, 17)

In the art world, a huge range of conceptual roles for the model has been explored. Particularly relevant, because it comes close to that of product designers, is the work of Gregory Green who builds models of bombs (figure 5.3), technological objects that look as though they work but do not. Although they could be made to work, their interest stems from the fact that the knowledge embodied in them is widely available and very destructive. The integration of the "bombs" as booby traps into familiar objects like suitcases links them to the world outside the gallery. Their technical uselessness becomes part of their value, shifting attention to their role as conceptual machines that engage the imagination and draw the viewer into a reflective and critical space.

These devices look similar to K/K Research and Development's analogs assembled from found machine parts. But these only work in relation to a narrative, usually social and political, in an accompanying text. They engage the viewer but are not powerful in themselves.

Crib-batic, another project by K/K Research and Development, in collaboration with Scholz, is a model for a push-chair, an existing object type. We know these objects exist in everyday life, how they are used, and by whom. So we can imagine what it would mean for their proposal to enter everyday life. It is not necessary to see the *Crib-batic* "working" for it to be effective, but rather to sense how it might cross over into everyday life. An imaginative alternative, it is

Figure 5.3 *Nuclear Device #2. 15 kilotons, plutonium 239* (1995) is a model, a technological object that looks as though it works but does not. Although it could be made to work, its interest stems from the fact that the knowledge embodied in these objects is widely available and very destructive.

"fact" in that it could be built, but fiction in that it is unlikely to be built. This fictiveness enables it to function critically, by highlighting the boundaries that limit everyday experience. It celebrates the complex ambiguity of the object, as both part and not part of the society from which it emerged. It has not acceded to the demands of "miserable reality" but remains defiantly conceptual.

From a product design point of view these models lack industrial realism; they look like craft objects, hand-made and probably one-off. But an expanded view of the conceptual design model might regard it as embodying the essence of a design idea, a "genotype"[2] rather than prototype, constructed from the materials at hand. If taken up for mass manufacture its construction and structure would undoubtedly change. The object's "content" or "genes" are important, not its appearance. In the context of design, the conceptual model as genotype rather than prototype could allow it to function more abstractly by deflecting attention from an aesthetics of construction to an aesthetics of use. The genotype depends on the view that a design idea can transcend its material and structural reality and function critically, in relation to social systems for example, rather than visual culture. Andrea Branzi (1984, 141) suggests this as a possible role for craft in late-twentieth-century industrialized production. Experimental furniture such as Studio Alchymia's 1980 *Bauhaus 2* range (figure 5.4) do not simulate how they would be if mass-produced, but take a form appropriate for exhibition and consumption as one- or two-offs. Rather than an

Figure 5.4 Andrea Branzi's *Ginger* (1980) for Studio Alchymia does not simulate how it would be if mass-produced, but takes a form appropriate for exhibition and consumption as one- or two-offs. The craft object is seen as a stage in the development of an idea that might eventually be mass-produced.

autonomous form of design, the craft object is seen as one stage in developing a design idea that might eventually be mass-produced.

Michele De Lucchi presented design studies for small domestic electric appliances (figure 5.5) at the 1979 Milan Triennale. They echoed a contemporary concern to challenge prevailing images of domestic technology. They are interesting because they do not mimic reality; they are clearly representations, "models" comfortable with their unreality. They are things in themselves rather than shadows of yet to be realized products. They offer real experiences of ideas rather than unreal experiences of unrealized products, and accept that these ideas will be consumed through books and exhibitions not in the marketplace.

The Design Object as Prop

By abandoning the technical realism of the prototype and the visual realism of the traditional industrial design model, conceptual models in combination with other media, can refer to broader contexts of use and inhabitation. For instance, by using conceptual models as film props the viewer can be drawn into the conceptual space of the object in use rather than an appreciation of the thing in itself.

Branzi suggests the age of the "Historical Avant-Garde" is ending. Large corporations work with small experimental design centers to develop new scenarios within which the corporations develop new products. He calls this a period of "Permanent Avant-Garde," the aim of which is "to restructure the market, to develop a new ecology of the natural and artificial environment, and to create islands of meaning that define consumption not as a category of the emphemeral and provisional, but as a solid culture for a democratic and reformed society, one in which a new generation of tools will be able to liberate people from uninspiring work, encouraging mass creativity and individual freedom" (Branzi 1995, 152).

A key tool in this process is the scenario, both to generate design ideas and communicate the results. Large corporations employ scenarios of use to anticipate how people will interact with the complex environments of which technological products are a part. Usually scenarios have a conservative role, predicting patterns of behavior in relation to technological developments. They draw from what we already know about people, and so weave new ideas into existing realities. These scenarios extend preexistent reality into the future and so reinforce the status quo rather than challenging it. For example, "Workshop" by Philips in collaboration with Olivetti explores the new office landscape to formulate a new vision of the workplace and propose new tools to support it. But the way it

Figure 5.5 Michele De Lucchi's *Appliances* for the 1979 Milan Triennale do not mimic reality; they are clearly representations, "models" comfortable with their unreality.

was conceived only reinforced stereotypes of the future office. *Communicator* (figure 5.6) for "anywhere anytime" multi-media computing, the group tool (figure 5.7) that encourages office workers to mix fundamental tasks such as photocopying and faxing with socializing, and the fetishistic arrangement of tools that can be interconnected to meet specific functional requirements propose no innovative vision of changing patterns of work. Corporations need to ensure a continued need for physical products in a world where many products are being replaced by software (e.g., phone and fax software for computers). But as a tool for presenting design ideas, the scenario is very powerful. It can draw the viewer into a narrative that goes beyond the object to reveal more complex issues.

Manzini (1994) argues that, although design can neither change the world nor create lifestyles that enforce patterns of behavior onto society, the designer is not simply a problem solver but an intellectual able to link "the possible with the hoped-for in visible form." Manzini's emphasis is less on interaction with discrete objects than on systems of objects. He suggests designers as independent agents use their imaginative skills to propagandize socially and politically desirable situations. In Manzini's view, part of the designer's role is democratically to discover what is "desirable" rather than imposing their own or another minority vision onto society. But Manzini's approach, although critical in that it rejects prevailing conditions and proposes an alternative, runs the risk of being either too didactic or utopian.

The sci-fi genre offers a third possibility. Susani, noting how what was once called "concept design" has now become the design of entire scenarios of objects, refers to Apple's 1987 "Knowledge Navigator" project as probably the first use of video narration to present a "cultural project" (Apple Computer, 1992). Susani claims it was neither a promotional tool, nor simply a projection of technological evolution, but a study of how we could coexist with new technological artifacts. He suggests that Wim Wenders' film *Until the End of the World* is a more stimulating and useful project for a "telephone scenario" than many mainstream design projects for telephones of the future. The use of scenarios in *Until the End of the World* comes close to being critical because it achieves a degree of estrangement through the behavior of fictional characters who do not have to conform to existing personality types, occupations, or motivations.

But this approach falls foul of a central contradiction of radical work, as Adorno demonstrated in his contrasting of modern classical music and popular jazz. Because a mainstream film has to be immediately graspable by a broad audience, the fact of achieving this diminishes its critical potential. Transformations of

Figures 5.6–5.7 Philips and Olivetti's *Communicator* (1994) and *Group Tool* (1994) set out to explore the new office landscape, to formulate a new vision of the workplace, and propose new tools to support it.

consciousness are more likely through struggling to understand ideas: simpli-
fication dilutes the power to challenge established values:

According to Marcuse, the strength of art lies in its Otherness, its incapacity for ready assimilation. If art comes too close to reality, if it strives too hard to be comprehensible,
accessible across all boundaries, it then runs the risk of becoming mundane. And if this
occurs, its function as negation to the existing world is abandoned. To be effective art must
exert the capacity for estrangement . . . it must dislocate the viewer, reader, audience, by
its refusal and inability to become part of the reality principle. (Becker 1994, 119)

If the conceptual design object is to be used as a prop in a scenario that works
in a critical, transformative way, other possibilities must be developed. Although a critical approach might alienate some, it might also more effectively
transform the consciousness of those whom it does engage. The task is to embody content in an aesthetically challenging form that would "push the viewer
towards a more complex, emotional, or revolutionary understanding of the
problems posed by the work" (Becker 1994, 122).

Some artists and sculptors have achieved this in films they have made about
their work (e.g., Philippe Ramette, or Rebecca Horn), and filmmakers such
as David Lynch have developed strategies for applying this to television (e.g.,
Lynch's *Twin Peaks*). But there are few examples (one being Atom Egoyan's *Family Viewing*) where electronic products play a significant role.

Cindy Sherman's photographs from her *Untitled Film Stills* series portray banal moments of apparently little significance. As the viewer tries to imagine
what happened before and after, he or she is drawn into speculation on the psychology of the protagonists and their state of mind. These photos show the surprising power of stills, compared to video or film, to engage the viewer. They
shift the viewer's imagination toward the fictional possibilities of the portrayed
moment. The furnishings and incidental objects in these photographs encourage an allegorical reading that further engages the viewer. Most of the images
look as though they were taken in the 1950s or 1960s which adds to the distance they create.

The work of Garner (1983, 1985) also uses a sense of the recent past to engage audiences. His two books consist of photographs (figure 5.8) of conceptual
design objects, and of scenes reflecting the strange psychological and social narratives that arise from interaction with and through the objects portrayed. The
books could be seen as a critique of consumer society, but their dependence on
comic absurdity distracts attention from any serious criticisms that might be

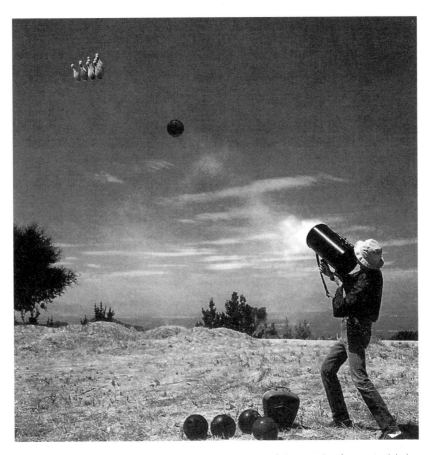

Figure 5.8 Philip Garner's *Utopia or Bust* (1985) consists of photographs of conceptual design objects and of scenes that present unusual narratives that arise from interaction with and through the objects portrayed.

read into the project. This ironic approach offers no constructive suggestions. In comparison, Ito's *Dwelling for a Tokyo Nomad Woman,* an architectural fiction conveyed through photographs, portrays a system of behavior and consumption to make familiar but exaggerated consumer values real and concrete, values that are neither futuristic nor utopian, but uncomfortably close to our own. The nomad woman's only furniture is designed to support intelligence gathering on new trends, eating snack food, and styling one's image. Ito's photographs conjure up an "elsewhere," familiar but different. Rather than offering another option, or parodying what exists, they suggest that the way things are is not the only possibility.

Although far more nostalgic and romantic, the images produced by Ramette, of himself using his inventions, work in a similar way (figure 5.9). The style of his images is deliberately straightforward, and the use of his devices, which usually resemble nineteenth-century scientific instruments, is easy to understand. The viewer wonders at the strangeness of Ramette's behavior, trying to imagine why somebody would behave like this, what pleasure they have, and what prevents such objects from being widely disseminated and the values they embody gaining general acceptance.

In Horn's films, *Der Eintanzer* (1978) and *La Ferdinanda: Sonata for a Medici Villa* (1981), her sculptures appear in the background of several scenes. They are never explained, but the viewer is drawn into a strange world that objects such as these seem to inhabit nonchalantly. The films seem set in the present, but the integration of such strange objects into everyday settings implies a completely different set of cultural and aesthetic values highlighted by their familiar settings. This technique is reminiscent of Brechtian alienation, in this case drawing our attention to the role of objects in defining and realizing everyday space and rituals. Horn's films are neither didactic nor utopian, nor are they parodies. They seem closer to heterotopias. They portray situations different from our own where enchanted objects have a place in daily life and a different "sense" prevails, a sense interwoven with our own rather than completely alternative or nonsensical. Norman Daly's *The Lost Civilisation of Llhuros* is an exhibition of artifacts from a fictional culture, each of which is accompanied by a caption explaining what is supposedly known about it. The exhibition blurs the boundaries between imaginary spaces and the here-and-now of the gallery. It is as though a film has reentered everyday life through its props. It invites the visitor to speculate, as an anthropologist of material culture might, on how values come to be embodied in artifacts.

Conceptual Consumerism

For Marcuse, art is a location—a designated imaginative space where freedom is experienced. At times, it is a physical entity, a site—a painting on the wall, an installation on the floor, an event chiselled in space and/or time, a performance, a dance, a video, a film. But it is also a psychic location—a place in the mind where one allows a recombination of experiences, a suspension of the rules that govern daily life, a denial of gravity. It "challenges the monopoly of the established reality" by creating "fictitious worlds" in which one can see mirrored that range of human emotion and experience that does not

Figure 5.9   The style of Philippe Ramette's images is deliberately straightforward, and the use of his devices, which usually resemble nineteenth-century scientific instruments, is easy to understand.

find an outlet in the present reality. In this sense the fabricated world becomes "more real than reality itself." Art presents the possibility of a fulfilment, which only a transformed society could offer."
—C. BECKER, "HERBERT MARCUSE AND THE SUBVERSIVE POTENTIAL OF ART"

This chapter has discussed where this space might lie in relation to the electronic as conceptual design object, and how we might encounter it. As a route for developing critical electronic objects within a design context, it has rejected the prototype in favor of combining nonworking models with film, video or photography to establish scenarios that are neither didactic nor utopian but heterotopian. Were the props from a scenario physically displayed with the film, video, or photograph, more subtle interactions might develop between the space of the here and now, where the viewer is, and the fictional space portrayed in the image. The physical presence of the artifacts encourages additional interplay between reality and fiction, between what is and what might be. By themselves the artifacts would be mentally assimilated into known patterns of behavior, "explained away." But shown as part of an alien culture with different aesthetic values and a different "sense," they require viewers to accommodate the unusual role of the artifacts in an everyday life like their own.

The space in which the artifacts are shown becomes a "showroom" rather than a gallery, encouraging a form of conceptual consumerism via critical "advertisements" and "products." New ideas are tried out in the imagination of visitors, who are encouraged to draw on their already well-developed skills as window shopper and high-street showroom frequenter. The designer becomes an applied conceptual artist, socializing art practice by moving it into a larger and more accessible context while retaining its potential to provoke people to reflect on the way electronic products shape their experience of everyday life.

6

Hertzian Space

The rapid expansion of knowledge and technical development has swept us into a world beyond our grasp; the face of nature is alien once again. Like the forest and the mountains of medieval times, our new environment harbours strange menacing beasts, invisible viruses, atoms, mesons, protons, cosmic rays, supersonic waves.
—GYORGY KEPES

It might seem strange to write about radio,[1] a long-established medium, when discussion today centers on cyberspace, virtual reality, networks, smart materials and other electronic technologies. But radio, meaning part of the electromagnetic spectrum (figure 6.1), is fundamental to electronics. Objects not only "dematerialize" into software in response to miniaturization and replacement by services, but literally dematerialize into radiation. All electronic products are hybrids of radiation and matter. This chapter does not discuss making the invisible visible, or visualizing radio, but explores the links between the material and immaterial that lead to new aesthetic possibilities for life in an electromagnetic environment. Whereas cyberspace is a metaphor that spatializes what happens in computers distributed around the world, radio space is actual and physical, even though our senses detect only a tiny part of it.

It is just over one hundred years since electricity generation started, seventy since radio transmissions began, and fifty since radar and telecommunications entered our environment. The twentieth century has seen space evolve into a complex soup of electromagnetic radiation. The extrasensory parts of the electromagnetic spectrum form more and more of our artifactual environment, yet

Figure 6.1 Electromagnetic Spectrum showing the Radio Frequency Spectrum.

designers direct little attention toward the possible sensual and poetic experience of this industrially produced new materiality.[2] In design, immateriality is often referred to through visual motifs, usually in relation to product semantics and representation, but it is rarely dealt with directly as a physical phenomenon.

Tuneable Reality

The extrasensory nature of electromagnetic radiation often leads to its treatment as something conceptual—which easily becomes confused with the notional, although of course it is physical and exists in space. The conflict between the conceptual and the perceptual aspects of hertzian space is an appropriate vehicle for investigating the boundaries between the imaginary and the actual.

Ray Lee and Harry Dawes exploit this ambiguity in *In the Ether,* a combination of film, theremin music, and performance, which takes the audience into a nostalgic and surreal realm between fiction and actuality: "They like it up high, radio waves. If you attach a long piece of wire to a long pole and put it up high you can hear them. I imagine that they spend all their time racing each other around and around the earth. I don't expect they come down very often. Except if they are curious" (Lee and Dawes 1996). Their film, made with Frances Boyle, taps into vaguely paranormal myths of radio folklore concerned with the mysteries of magnetism, and offers what can be described as a "psycho-hertzian" reading of everyday life.

Aerial Paris and *Aero-Living Laboratories* by Lebbeus Woods, a re-siting of architecture in electromagnetism, exploit this ambiguity less successfully. Electromagnetism becomes a field that "binds building to the sky instead of the earth."[3] Although one of the few architectural propositions centred on the electromagnetic aspects of space, this "architecture suspended in an invisible matrix of air and charge" is a form of science fiction. Its grand speculations and escapist logic cannot match the gently provocative poetry of Lee and Dawes. It is difficult to see exactly what Woods's two projects gain through their association with electromagnetism in terms of architecture or new models of living

Another architect, Laura Kurgan, responds more directly to inhabiting a ubiquitous electrosphere projected onto earth by a network of satellites.[4] Using what Virilio calls "the little everyday object [that] probably constitutes the event of the decade as far as globalisation of location goes," the GPS (Global Positioning System) navigator, she rigorously maps her explorations of this space. The GPS, which uses military satellites to plot the position of a sensor anywhere on the planet, is currently not very accurate, partly because for security reasons the military does not want civilians to have access to such accuracy and partly because the signal is distorted through reflections near the sensor (e.g., by buildings). Kurgan uses the GPS to map a space somewhere among the physical, digital, and conceptual. She stands in a gallery stationary for ten minutes recording 311 position records, plots the results on a map of the gallery and its surroundings and compares them with a more accurate computer corrected version.[5]

The artist Ingo Günther's site casting describes a situation where the television signal does not travel to where you are; the reverse happens. You have to go where it is; you have to hunt for it. He imagines a city of tiny television transmitters broadcasting the forgotten pasts of buildings, places, and streets. The city becomes a tuneable urban environment. Different time periods could be arranged as different channels into which the participant could tune. To do this it is necessary to design aerials more directional than usual, allowing different signals to be spatially separate in one location. The resultant antennas resemble the sculptures of Klaus vom Bruch who as early as 1984 exhibited constructivist-inspired aerials for broadcasting video signals between elements of his installations. For Günther's video installation, *Exhibition on Air,* the visitor enters the P3 art and environment gallery in Tokyo and wanders about holding a combination of aerial and LCD television, receiving broadcasts from other antennas sited around the building.

A different kind of narrative space is explored by Scanner, who uses a wideband radio scanner to tune into cellular telephone conversations, combining them on CDs to create ambient and often poignant sound images of the psychological and social poetry of everyday radio space.

These urban analyses of the militarization of the spectrum, tuneable urban narratives, and audio snapshots of telephone "normality" contrast with the New

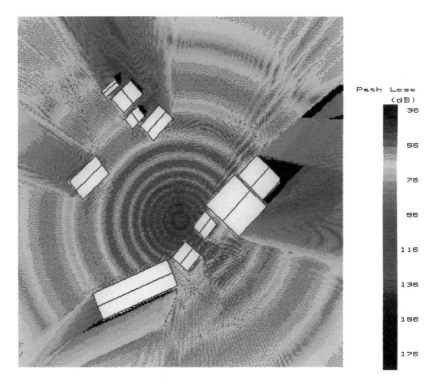

Path Loss
(dB)

35

55

75

95

115

135

155

175

Figure 6.2   (see opposite)

Age mysticism of sculptor Michael Heivly who, like Woods, seems to encourage an escape into the fantastic rather than a confrontation with and possible transformation of existing reality. Heivly writes about how microwave form, detectable yet unavailable to the senses,[6] represents simultaneously the idea of the real and the imagined. His work translates landscapes into musical sound compositions transmitted as microwaves into deep space. The microwave energy becomes a cone-shaped sculptural form that moves through space at the speed of light and retains its form for millennia. All his work aims to create an environment that confronts participants with the known and the unknown, and requires them to use their imaginations to construct the piece in deep space.

Electroclimates

Computer models showing radio propagation in relation to urban environments (figure 6.2), and maps showing the field strength and "footprint" of television and radio transmissions in relation to the surface of the earth (figure 6.3), reveal that hertzian space is not isotropic but has an "electroclimate" defined by wave-

Figures 6.2–6.3 Computer-generated models showing radio propagation in relation to urban environments, and maps showing the field strength and "footprint" of television and radio transmissions in relation to the surface of the earth, reveal that hertzian space is not isotropic but has an "electroclimate" defined by wavelength, frequency, and field strength arising from interaction with the natural and artificial landscape.

length, frequency, and field strength arising from interaction with the natural and artificial landscape.

The extent of the electrosphere is reflected in the difficulty of finding electromagnetically unpolluted parts of the globe as sites for intelligence gathering, "antenna farms," the use of faraday cages to create "empty" zero-field spaces for isolating sensitive equipment, and the realization that a modern war is won by the side that best exploits the electromagnetic spectrum by denying the enemy its effective use and protecting friendly electromagnetic systems against electronic attack:

Before we bipedal apes invented radio receivers, before we even exchanged our gills for lungs, there was radio. It was in lightning, in hydrogen atoms, in the big bang that propelled our universe into existence. But as soon as we invented technology that enabled us to listen to the transmissions of our planet, we saturated the airwaves with our own sounds—garage door openers, cordless phones, baby monitors, police dispatchers, pagers, and wireless microphones—jamming the oldest radio station around. (Strauss 1993, 336)

"Whistler hunters," natural radio enthusiasts who search out radio transmissions created by atmospheric events, map the interface between atmospheric and electromagnetic climates. They search out natural radio signals, VLF (very low frequency) radio waves or "sferics" (short for atmospherics: natural radio-frequency emissions in the ionosphere, caused by electromagnetic energy radiated from lightning). These signals—resonant clicks and pops called "tweaks" and "bonks" by scientists—occur in the audible range and may be picked up by antennas and amplified for listening. They are best received at night, far from power lines. Occasionally sferics get caught on, and travel long distances along, the magnetic flux lines around the earth, producing "whistlers," downward-gliding signals that may last up to three seconds. Whistler hunters travel far from power lines and electromagnetic pollution, sometimes camping out for days, listening for the elusive sounds of natural radio.

Between 1967 and 1975, the composer Alvin Lucier became interested in these sounds and made performances using prerecorded whistlers. In 1981 he recorded whistlers and spliced together short samples in chronological order for Sferics (1988). Other variations were produced for performances during the early 1980s, one of which involved setting up a small array of antennas at a campsite for the public to listen to in real time through battery-powered tape recorders and headsets. But it is doubtful that such artificial events capture the poetry of the whistler hunters' activities. Although the sounds are fleetingly beautiful, out of context they lose much; their beauty is entwined with the effort endured and the symbolic significance of receiving them, which for some is quasi-mystical, for others a defiant gesture against people's careless attitude toward nature.

More successful if less romantic celebrations of the electroclimate of artificial radio have been achieved through radios used as performing instruments by other composers. This began with Cage's "Imaginary Landscape No. 4 for 12" receiving sets: the arbitrary nature of broadcast material must have appealed to the composer of the "Music of Changes" and "4'3"." Later, in "Kurzwellen (Shortwaves)," Stockhausen used radio sounds to open himself to a "music of the whole earth":

What can be more world-wide . . . more ego-transcending, more all embracing, more universal and more momentous than the broadcasts which in Kurzwellen take on the guise of musical material? . . . What happens consists only of what the world is broadcasting now; it issues from the human spirit, is further moulded and continually transformed by the mutual interference to which all emissions are subject; and finally it is brought to a higher unity by our musicians in their performance." (Griffiths 1986, 165–166)

Whereas these composers celebrated the ubiquity of hertzian space, for Architekturbüro Bolles + Wilson the electronic glare of an invisible ephemeral city of ubiquitous impulses is something to be sheltered from, where comfort is to negate for a moment this network to create a zone of electronic shadows. Their proposal is less about the poetics of revealing the world as it is, and more about charging architectural space with psychological dimensions derived from acknowledging hertzian space.

Immaterial Sensuality

We are experiencing a new kind of connection to our artifactual environment. The electronic object is spread over many frequencies of the electromagnetic spectrum, partly visible, partly not. Sense organs function as transducers, converting environmental energy into neural signals. Our sense organs cannot transduce radio waves or other wavelengths outside the narrow bandwidth of visible light (and infrared energy through the skin as warmth). Electronic objects are disembodied machines with extended invisible skins everywhere. They couple and decouple with our bodies without us knowing. Working on microscopic scales, often pathogenic, many electromagnetic fields interfere with the cellular structure of the body. Paranoia accompanies dealings with such hertzian machines. How do they touch us? Do they merely reflect off our skin, or the surface of our internal organs? In other words, do they merely "see" us, or can they "read" us too, extracting personal information about our identity, status, and health?

An operating manual for X-ray machines contains images of radiographic actors and props (figure 6.4) that view the body as a radio medium. The machines establish views, and support a sort of radio perspective, revealing, concealing, and exposing hidden organs and views, and creating a "radio theater" of the hidden body. In configuring the body according to an unusual conception of space, these images of people and X-ray machines illustrate an expanded view of space as an electromagnetic medium.

Figure 6.4 An Ilford manual for X-ray machines contains images of radiographic actors and props that illustrate the use of the body as a radio medium.

The artist Arthur Elsenaar, inspired by photographs of experiments by gentleman scholars in the 1850s, taps into our strong feelings about electricity, its danger and mystery, and its measurability. He uses the microwave field of a radar sensor to create "an aura, or an extension of my skin into spaces, into which people can walk," which causes a 24-volt pulsed DC current to deliver a variable charge of up to three milliamps to two electrodes attached above the jaw and two to the "hunch" muscles in the shoulders. He is developing a digital system that will support a wider range of inputs and outputs—for example, different responses for people retreating and approaching, and head turns and nods.

Elsenaar's poetic use of fields, a pathological exploration of personal space, is very different from research carried out at MIT into technically interesting but aesthetically mundane applications of electrical fields to interpersonal informa-tion exchange. The body is treated literally as a circuit board and the commer-cial ambitions of the project have eradicated any possibility for poetry although the "electrical whispers of fish" are one inspiration for the project.[7] The poten-tial of the technology is reduced to the most basic level of utility and conceived as a replacement for physical connections between personal databases.

Making Visible the Invisible

Long before radio energy was used to carry an acoustic signal, many ingenious devices were invented to detect radio energy[8]—for example, Lefeuvre's "physiological" receiver (figure 6.5) that uses the electrical sensitivity of the frog's leg. These objects resemble the early meteorological equipment used to make visible atmospheric phenomena otherwise too subtle for our bodies to sense. Just as the barometer tells us how heavy the sky is, these early radio detectors embody a more poetic understanding of hertzian space by revealing the extent of its presence.

Today a typical urban radio frequency environment is dominated by radio and television broadcast transmissions. Other forms of radiation such as microwave relay links, radio telephones, CB, speed-detecting radar, satellite communications systems, military tracking radar, civilian air traffic control, air route surveillance, and weather radar all combine into what Ito has called "Active Air."[9]

Several of Ito's works evoke this implied sensuality. In his *Tower of Winds* (figure 6.6)—realized in 1986 in the middle of a neon downtown, in front of Yokohama station—he wanted the air itself to be converted into light. The tower appears to dematerialize at a particular moment, reappearing in response to ambient noise levels. His *Dreams Room* (figure 6.7), installed at the Victoria and Albert Museum in London in 1991–1992, tries to evoke the immaterial sensuality of the new information environment by combining an information-saturated environment of projected imagery with specially commissioned interface objects,[10] intended to reinforce at an intimate scale what the environment communicates at the scale of architecture.

Although at first sight *Signals* by Takis looks like antennas responding to the contents of the air, low-tech precursors of Ito's *Tower of Winds,* they are in fact nonworking symbolic evocations. His work is important not only because it dealt with the poetry of electromagnetism in the 1950s long before others, but also because he developed a language that referred indirectly to the mysterious and metaphysical aspects of electricity and magnetism, in contrast to the more exuberant responses to technology of many of his contemporaries.

Ito's and Takis's pieces visually imply they are translating the invisible into the visible. Another piece of architecture, *House under High Voltage Lines* (figure 6.8) by Kazuo Shinohara, at the other end of the technological scale from Ito, provides an equally beautiful but more restrained response to the new technological situation brought about by electromagnetic space. The site is beneath high-voltage power lines. Strict regulations determine the safe distance from these lines, and the roof of the house defines this zone for two cables, creating an interface between a possibly pathogenic electromagnetic field and a sculpted

Figure 6.5   Long before radio energy was used to carry an acoustic signal, many ingenious devices were invented to detect radio energy. Lefeuvre's "physiological" receiver, for instance, uses the electrical sensitivity of a frog's leg.

interior space. It evokes more disturbing and powerful notions of radio space than the work of either Ito or Takis because it exposes the possible harmfulness of these fields and their existence in everyday life.

The architects Herzog & De Meuron respond to an electromagnetic context in their design for a signal box (figure 6.9). The building houses sensitive electronic equipment and needs to be protected from sudden bursts of electromagnetic radiation. It is not clear if this proposal is an alternative or an augmentation of the usual techniques of shielding, but it has resulted in a powerful image of architecture situated within hertzian space. Although relatively low-tech and programmatically mundane, this is another example of how sensual material responses to immaterial electromagnetic fields can lead to new aesthetic possibilities for architecture. It contrasts with more self-conscious rhetorical expressions of electronic culture by architects such as Jean Nouvel, Rem Koolhaas, and Bernard Tschumi.

The Radiogenic Object

Objects designed to straddle both material and immaterial domains arouse curiosity about the fit between these worlds. Many military aircraft are now "teledynamic," designed to fly undetected through fields of radar-frequency radiation. But teledynamic forms are not aerodynamic and to remain airborne their outline needs to be constantly adjusted by a computer. These aircraft fly through fusions of abstract digital, hertzian, and atmospheric spaces. If this awareness of hertzian space is to form the basis of an approach to everyday objects, it is not enough simply to present the technical facts. They must be grounded in rich cultural contexts if they are to be more than mere illustrations.

Objects that I call "radiogenic" function as unwitting interfaces between the abstract space of electromagnetism and the material cultures of everyday life, revealing unexpected points of contact between them. Many of these objects centre on the aerial, a device that links the perceptible material world to the extrasensory world of radiation and energy.

"Aerialness" is a quality of an object considered in relation to the electromagnetic environment. Even the human body is a crude monopole aerial. Although in theory precise laws govern the geometry of aerials, in reality it is a black art, a fusion of the macro world of perception and the imperceptible world of micro-electronics. Embodying the contradictions and limits of scientific thought, an aerial's behaviour can be described but not easily understood because it depends on the dual concept of electromagnetic radiation as wave and particle. As the aerial allows this invisible world to be understood and modeled in terms of material reality, it provides a starting point for a design approach that links the immaterial and the material so as to open up new aesthetic and conceptual possibilities.

Although few artists have explored radiogenic objects, several objects have been created by radio amateurs by enhancing radiogenic qualities in existing environments and artifacts, resulting in objects that provide new perceptions of our hertzian environment. These objects hint at the fertile territory beyond the designer's concern with the semiotics of radio interfaces and the engineer's narrow conception of functionality.

Do-it-yourself (DIY) books on antenna theory and practice offer many examples that generate the kind of pleasure that Wentworth shows in *Making Do and Getting By* and people's natural ability to subvert object types and act in new

Figure 6.6 Several of Ito's works attempt to evoke an implied sensuality. In his *Tower of Winds* (1986)—realized in the middle of a neon downtown, in front of Yokohama station—he wanted the "air itself to be converted into light."

ways on the environment. It is a pleasure derived from invention as poetry, loosening the connection between language and things, and challenging the tyranny of language over artifacts. For instance, the use of a specific size of domestic shelving as a core antenna (figure 6.10) or a milk bottle and tin foil for another (figure 6.11) reveal unexpected functional connections between physical objects and hertzian space that offer an alternative to representation.

Another example is the bobbin cane (figure 6.12), conceived and made by Georges Droz-Georget for listening to the forbidden French transmitter on the Eiffel Tower during World War I. A hook passing through the hole of the ferrule was attached to a low overhead telephone line connected to a shooting

Figure 6.7 Toyo Ito's *Dreams Room* (1991–1992) for the Visions of Japan exhibition at the Victoria and Albert Museum in London evokes the immaterial sensuality of the new information environment.

range: this became his antenna. The wavelength was selected by moving two sliding rings over the copper thread wound around the shaft, and the receiver was carried in the user's pocket. All these radiogenic objects are part of a hertzian culture that includes diagrams on the use of drain pipes as antennas, and garden layouts (figure 6.13) that integrate an antenna with vegetables and paths.

When Objects Dream . . .

Although when we look at an electronic product we only see what is radiated at the frequency of visible light, all electronic objects are a form of radio. If our eyes could see (tune into) energy of a lower frequency these objects would not only appear different but their boundaries would extend much further into space, interpenetrating other objects considered discrete at the frequency of light. Besides the obvious harmfulness of X-rays and microwaves there is a growing concern over the effect of the radiation leaked by domestic appliances. *Radio and*

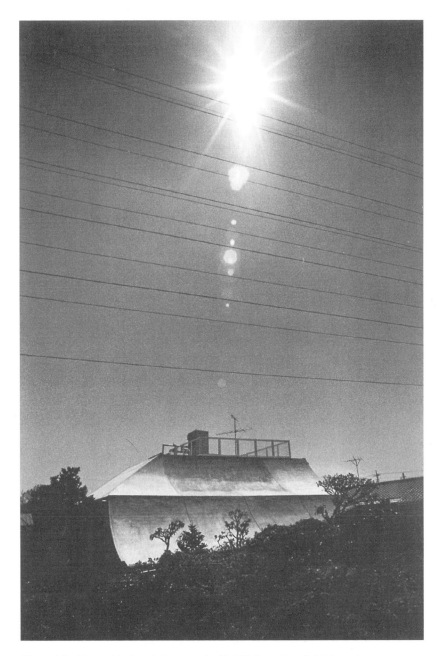

Figure 6.8 Kazuo Shinohara's *House under High Voltage Lines* (1981) provides a conceptually eloquent response to the new technological situation brought about by electromagnetic space.

Figure 6.9 Herzog & De Meuron's *Signal Box #4* (1991–1994) is an example of how sensual material responses to immaterial electromagnetic fields can lead to new aesthetic possibilities for architecture situated within hertzian space.

Beans, an installation by Patrick Ready, draws attention to the possibly harmful effects of electromagnetic fields generated by domestic appliances. It consists of electrical devices suspended on wooden shelves from the gallery ceiling. Around them hang small paper bags containing fast-growing mung beans in soil, arranged at equal intervals in a three-dimensional grid and watered three times a day. It was hoped that the beans would exhibit effects from the electrical fields through irregular growth patterns. But as the experiment was not controlled and scientific but ironic, it was never clear how the beans were affected.

In this piece the artist becomes a radio biologist investigating the interaction between radiant energy and biological systems. Science and folklore meet in this strange electrical garden, reminding us of the interconnectedness of nature and technology, something that must be made more visible if we are to find more meaningful ways of inhabiting an environment gradually becoming more radioactive.

The electronic object is often described as "smart." But using this term to describe objects with enhanced electronic functionality encourages a bland interpretation of electronic objects: "Smart, after all, is not the same as intelligent, let alone intellectual. Smartness is intelligence that is cost-efficient, planner-responsible, user-friendly, and unerringly obedient to its programmer's designs. None of the qualities, in other words, which we associate with free-thinking intellectuals" (Ross 1994, 331).

Electronic objects are not only "smart," they "dream"—in the sense that they leak radiation into the space and objects surrounding them, including our bodies. Despite the images of control and efficiency conveyed through a beige visual language of intelligibility and smartness, electronic objects, it might be imagined, are irrational—or at least allow their thoughts to wander. Thinking of them in terms of dreaminess rather than smartness opens them to more interesting interpretations.

For example, some possibilities for new relationships with these hybrids of radiation and matter are found in pathological products based on paranoia or eccentricity. Many devices designed to transform private situations into public ones depend on the "leakiness" of electronic objects, tuning into the dreams of radiant objects. The Computer Intercept System sold by the Surveillance Technology Group (n.d.) is an example: "Without entering the premises, electromagnetic radiating from unshielded computer screens and ancillary equipment can be intercepted from a remote location. The Computer Intercept System's highly sensitive receiver logs all radiating signals into its 100 channel memory. These emissions are then stabilised, processed and reassembled into clear reproduction of the intercepted data onto its built-in monitor" (27).

Many buildings are now designed as faraday cages to prevent such eavesdropping, usually invisibly by deploying electromagnetic shielding materials throughout the structure. The same technology protects sensitive equipment in a building from bursts of external radiation. Test-sites, specially designed environments, or anechoic chambers now measure an object's leakiness to predict its effect on other objects.

A more bizarre use of leakiness is seen in the Bat Band Converter (figure 6.14), a parasitical device that allows you to "use your a.m. portable radio and this novel design to tune-in to the secret world of bats." The title of the magazine that provides the plans, *Everyday Practical Electronics,* seems at odds with a world where

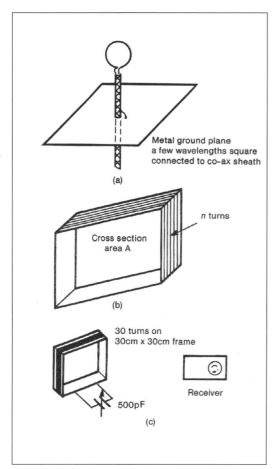

Metal ground plane
a few wavelengths square
connected to co-ax sheath

(a)

n turns

Cross section
area A

(b)

30 turns on
30cm x 30cm frame

Receiver

500pF

(c)

Existing ferrite antenna

Convenient furniture as frame

(d)

Insulating
gap

Pancake winding

Conducting
tube

Coil
leads

(e)

400
100
40
F
10
4
1
    1   4 10 40 100   Rod length/diameter

Multiplying factor F
for ferrite core
with permeability ~500

(f)

Figure 6.10   (see opposite)

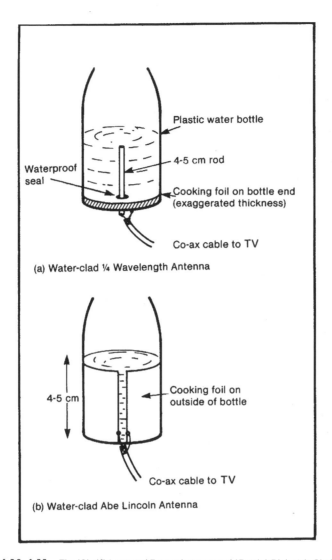

Plastic water bottle

4-5 cm rod

Waterproof seal

Cooking foil on bottle end (exaggerated thickness)

Co-ax cable to TV

(a) Water-clad ¼ Wavelength Antenna

4-5 cm

Cooking foil on outside of bottle

Co-ax cable to TV

(b) Water-clad Abe Lincoln Antenna

Figures 6.10–6.11   The (Shelf) Loop and Frame Antenna and (Bottle) Dielectric Clad Antenna are examples from DIY books on antenna theory. They generate the kind of pleasure associated with making do and getting by and with people's ability to subvert object types and act in new ways on the environment.

Figure 6.12 Droz-Georget's bobbin cane was made for listening to the forbidden French transmitter on the Eiffel Tower during World War I.

Figure 6.13 Design layout for a garden allowing for the inoffensive integration of an antenna with vegetables and paths.

Figure 6.14 The Bat Band Converter from *Everyday Practical Electronics* is a parasitical device that allows you to "use your AM portable radio and this novel design to tune-in to the secret world of bats."

practical skills are turned toward poetic ends, and tuning in to bats, hair, fizzy drinks, crinkly plastic bags, and dropping pins is regarded as a sane everyday activity. The device converts the non-electromagnetic ultrasonic signals of the bats into radio signals that are transmitted/leaked to the host radio.

The seemingly illicit information exchange of "dreamy objects" offers one possible interpretation of the electrosphere. It helps us think of electronic objects in "hertzian" terms, as interconnected fields rather than discrete things. It acknowledges the problematic conceptual status of electronic objects, arising from their ambiguous identity as hybrids of matter and radiation, functioning at scales and speeds well beyond the range of human perception. If the electronic object has a role in humanizing hertzian space it is not as a visualization or representation of radio but as a catalyst, encouraging the poetic and multilayered coupling of electromagnetic and material elements to produce new levels of cultural complexity.

Hertzian Tales and Sublime Gadgets

The [Manufacture d'Armes de Saint-Etienne] catalogue itself, however—its actual existence—is rich in meaning: its exhaustive nomenclatural aims have the resounding cultural implication that access to objects may be obtained only via the pages of a catalogue which may be leafed through "for the pleasure of it," as one might a great manual, a book of tales, a menu.

—J. BAUDRILLARD, *THE SYSTEM OF OBJECTS*

This chapter includes a commentary on five conceptual design proposals for post-optimal electronic objects: Electroclimates, When Objects Dream . . . , Thief of Affections, Tuneable Cities, and Faraday Chair.

Each proposal is a material tale, a process of investigation. They are "value-fictions": they try to maintain a degree of technological realism while exploring values different from those current. Their subject is the role of electronic objects in the aesthetic inhabitation of a rapidly dematerializing, ubiquitous, and intelligent environment. They explore ways of presenting conceptual designs as investigations and processes rather than as finite things in themselves. Each proposal is a radio, an interface between the electromagnetic environment of hertzian space and people. Each explores different forms of realism: technical, functional, social, and psychological.

The proposals are not intended for mass production or even prototyping, but for mass consumption through publication and exhibition. They ask questions rather than provide answers and should stimulate discussion in the way a film or novel might. Each focuses on different design issues. "Thief of Affections" explores designing role models and psychosocial narratives. From "Electroclimates" emerges ideas for genotypes, pseudo-interviews, and poetic products. "When Objects Dream . . ." offers alternative conceptions of the smart object as dreamy object, and new tools like the gaussmeter for mapping hertzian space. "Tuneable Cities" explores overlapping electromagnetic and urban spaces using a car and scanner to experience a city. Architectural models emphasize radio as environment rather than medium, and video stresses the design of experience rather than that of objects. "Faraday Chair" investigates a conceptual approach to the aesthetics of hertzian space and the object.

Electroclimates: Abstract Radio

This proposal developed from my desire to create a post-optimal object that answered aesthetic needs within a context of everyday life. It would be an aid for poetically inhabiting the electrosphere, a contemplative object revealing the hertzian nature of our environment.

I began the investigation with the realization that hertzian space is not isotropic but has its own electromagnetic "climate" that is related to an electro-geography defined by wavelength, frequency, and field strength, and it interacts with urban and natural environments as discussed in chapter 6.

To make visible minute atmospheric changes, antique meteorological devices such as barometers, hygrometers, and thermometers often use unusual means that reveal the sensuous materiality of space. For example, to indicate humidity, some hygrometers exploit the expansion of hair and skin when moisture is absorbed, and the transparent design of early barometers shows they measure the "weight of the sky." My object would be an electronic relative of the early radio detectors which also employed ingenious means of indicating the presence of radio waves (figure 7.1).

My original proposal was a radio that converted electroclimatic changes into abstract sounds using a wideband radio scanner. It would allow one to notice patterns and become familiar with the flow of activity within a particular area. On another level, Electroclimates is a response to the communications that invade domestic spaces. When a scanner is used in the privacy of the home to listen in on a telephone conversation outside it, the user is seen as the invader but, seen from another viewpoint, the radio signals from cellular telephones are invading the home. Electroclimates uses an aesthetic language to gently draw attention to this new and problematic interface between private and public space.

Figure 7.1 Concept sketch.

An opportunity to use LCD screens arose. The first reaction was hesitancy. Screens are like "supermatter": once switched on, all attention turns to them, and their material qualities are demoted to the status of package or container as the viewer searches for the real content, information. Unlike sound, which can be nondirectional, screens tend to give a space a specific orientation.

To explore subtle and evocative uses for the monitor as a material, I found a way to use a screen to communicate gently and impressionistically, by experimenting with different plastics. When thin sheets of sanded fluorescent polycarbonate are held close to the screen they interact with its light to produce a hazy effect. I made a simple animation that slowly changed color and gently pulsated.

I then explored the physical nature of Electroclimates through rough sketches and scale models. By arranging the screen horizontally, I could view it from any direction, overcoming the dominance screens have on the layout of rooms. The two main areas of investigation were into Electroclimates as a piece of furniture such as a small table or a ceiling fixture, and a portable device.

I decided not to simulate industrial production but make a handmade geno-type, an object designed to communicate the essence of the idea that could later be developed for mass or batch production if the occasion arose. The entire object would be made from one material so that the screen appeared to dissolve into it.

As I experimented with the scanner and considered how Electroclimates would be used, one time of day became particularly interesting: late at night, as callers sleepily said good night to each other from their beds. This led to the idea of interacting with a product at the moment when boundaries between reality and dreams begin to merge. Electroclimates became a "pillow."

Electroclimates responds to local changes in the radio frequency environment by switching itself on when it detects signals stronger than the general background. It turns electrical space invasions of the home into flickering patterns of light and distorted sounds, when a head is placed on the pillow the distortion clears revealing what is actually being received (e.g., telephone conversation, fax transmissions, or garage door openers). Through a slow, gentle interaction, the owner would gradually learn to read his or her electromagnetic environment through the object's responses.

The final design consisted of an LCD screen encapsulated within a fluorescent polycarbonate box that is suspended in a clear PVC inflatable pillow (figure 7.2) and connected by a lead to a wideband discone aerial. Discone aerials

Figure 7.2 Final design for Electroclimates.

are usually located outside the home; this one is indoors to emphasise that radio-waves are penetrating domestic space.

The semi-working object was shown in the Monitor as Material exhibition at the Royal College of Art, which while demonstrating the potential of LCD screens, also provided an opportunity to test public receptivity to the idea of electronic products for answering poetic needs. When fully explained to visitors, Electroclimates elicited an enthusiastic response. Without an explanation, however, most people saw it only as an exhibition piece rather than a potential product.

As a result of feedback from exhibiting Electroclimates, I made a pseudo-documentary video in collaboration with Dan Sellars and Fiona Raby. An elderly woman in her home describes how she thought she would live with an object like Electroclimates, how she came by it, when she used it, and what she used it for. We explored where she would keep it, how often she would use it, and how her friends and neighbors might react (figures 7.3–7.6).

The intention was to steer between a number of established approaches: user-testing requires that the object works fully; product clinics test consumer reactions to a product based on how things are now, as are Design Age sessions with the University of the Third Age; "Informance" aims to persuade an audience

Figure 7.3   Still from *Pillow Talk* video.

Figures 7.4–7.5 Stills from *Pillow Talk* video.

Figure 7.6 Still from *Pillow Talk* video.

that a product fits in and has a place. But here the aim was not to convince an audience of a need, but to draw them into a "what if . . ." scenario, a "value-fiction" to stimulate a desire for change. The interviewee is a knowing participant in a fiction.

In some ways, Electroclimates "fails": it is too seductive to be a "critical design" in that the values it embodies are not strange enough.

When Objects Dream . . .

Most people are aware that products like desktop computers, faxes and televisions emit low level electromagnetic (EM) radiation, but it is still unclear if it is harmful. This proposal started as an object for electromagnetic spaces generated by electronic products.

It developed simultaneously along three different paths: ways of sensing and indicating the presence of fields, uses for registering the presence of fields, and the physical nature of the object itself.

My first idea was for small containers that avoided electromagnetic fields. They would be kept on a desktop and move away from the fields they detected. Another idea was to incorporate compasses into a tabletop so that fields from devices placed on it would become visible through the deflecting needles. But compasses are not sensitive enough to be influenced by radiation emitted by computers. Both ideas were for "enchanted" objects that would mysteriously come alive.

I looked at equipment for measuring VLF (very low frequency) and ELF (extremely low frequency) emissions from products, but these were beyond the budget of this project. So I used a gaussmeter a device for measuring the magnetic component of electromagnetic fields to measure and draw fields produced first by televisions and later by a computer, answerphone, printer, and fax machine arranged on a table (figure 7.7). The gaussmeter revealed an alternative vision of electronic objects as fields, which led to the idea of "dreamy objects" (see chapter 6). It was chosen as the technical basis of the project.

I explored more design ideas: adhesive nipples that vibrated when they sensed fields, warning the wearer to move back, seat backs with vibrating nodules that indicated radiation was passing through the sitter, and parasitical lights that only worked if positioned in fields emitted by domestic products.

Most of these ideas appeared either too whimsical or, in the case of the lights, too feasible. I returned to the idea of the electronic product as a dreamy object and decided to develop an "object for seeing the dreams of consumer products." (From this point on, When Objects Dream . . . was developed simultaneously with Electroclimates as part of the Monitor as Material exhibition.)

A "glove" was considered first, as though the wearer were caressing the invisible skin of the electronic product, locating its true limit. But this seemed too intimate: there should be more distance between the person and the dreams of products. I decided the device should only work when placed "at arms length" into the leaky field of an electronic object, a space we can never sense.

The frequency, wavelength, and intensity of the victim object's "dreams" influenced the color fields and sounds emitted by an LCD screen encased in fluorescent polycarbonate (figure 7.8). The final object was not made to look like an injection molding but to appear abstract and brutal. The part of the object touched by a person was made from a square block of wood, emphasising that the human qualities are not in the form but in what the device does. Two headphone sockets allowed the sounds to be shared.

Like Electroclimates, When Objects Dream . . . is a semi-working genotype designed for the Monitor as Material exhibition. Its screen shows a video of a computer animation. Ideally it would be presented juxtaposed with a consumer product like a television.

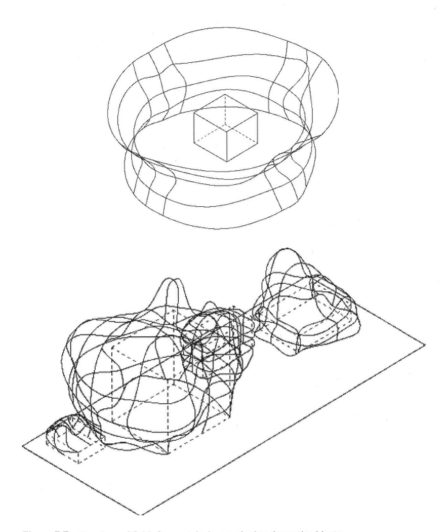

Figure 7.7 Drawings of fields from a televison and other domestic objects.

Figure 7.8 LCD screen encapsulated in fluorescent polycarbonate.

Thief of Affections

This proposal is based on the realization, discussed in chapter 4, that electronic products are "role models" and that when we use them we become the generic user they are modeled on.

Thief of Affections started with my desire to design an object that embodied an alternative model of a user, a "perverse" role model. This project is grounded in perversity: not sexual perversion but the desire to rebel, to deny the system the satisfaction of total conformism. Its use would place the user, now a protagonist, into a new relationship with the familiar, providing a new narrative dimension to everyday life.

Figure 7.9 Concept sketch.

The approach was similar to those architectural projects described as "a house for . . ." where a specific psychological model generates an unusual set of functional requirements. In this case the proposal was to be a walkman for an "otaku," a term used in Japan to describe an obsessive person, usually male, slightly perverse and socially dysfunctional. The design of the personality became part of the product.

My misreading of "Doppler Danse: Some Novel Applications of Radar" by Steve Mann (1992) led to the idea of using radar to caress the internal organs of unsuspecting strangers: the otaku, perversely attempting to experience intimacy by technologically groping the victim's heart (figure 7.9), would become a thief of affections. The caress would be converted into vaguely erotic sounds.

The project began to follow two lines of investigation: a technological investigation of the "caress" and how "affection" could be stolen, and an exploration of the physical nature of the "walkman."

At first I mistakenly assumed that different frequencies of radar penetrate the body to different depths rather than reflecting off its surface. I considered ultrasonic scanners, which do penetrate the body to different degrees, but they require the transducer to be placed in contact with a gel spread on the skin. When I discovered that the body gives off a very weak electromagnetic field, the idea of the thief stealing very weak radio emissions from the body appeared feasible. But these signals are so weak that highly specialized and bulky equipment would have to be used, and at very close range. And the technology had to be believable if the proposal was to be a "value-fiction" not a "science fiction."

I then considered the physical nature of the device. It could easily become a black box, an alternative walkman, or be incorporated into existing objects like clothing or the body (as an electronic tattoo for instance). Although it could be argued that the experience produced by the electronic technology alone is the product, I felt that the nature of the object itself was as important. The juxtaposition of experience and object counted in the sense discussed in chapter 1.

To generate the object I looked again at the character of the thief. Being "perverse," the thief would resist the trend towards miniaturization and absorption of electronics into existing objects or the body. Hypersensitive to technology penetrating his own body, he would favor glasses over contact lenses, baggage and walking sticks over pockets. The final object would be separate from the body, like a walking stick. The physical configuration of the object was then explored through sketches.

Considering how close technology can come to the body before it becomes invasive led to the pacemaker, the ultimate technological invasion of the body, which transmits weak radio signals. The thief would steal the radiation given off by the artificial heartbeat of a radio heart, becoming the "Thief of (Radio) Affections," placing himself in a new form of intimacy with his unsuspecting victims (figure 7.10).

The weak radio signals emitted by the pacemaker could be picked up by a test-probe of the kind normally used to measure stray electromagnetic emissions from domestic appliances. Although relatively expensive they could form a technical basis for the device. Essentially it would be a radio tuned to a very narrow part of the electromagnetic spectrum.

I then started exploring the physical design through scale models. I decided it should be audacious. The thief would display the fact he was engaged in some unconventional activity, but simultaneously provide himself with a conspicuously voyeuristic mask. The object became more decadent, like a riding crop. Originally conceived as being made from neutral materials it was now to be made of leather, referring to both its status as luggage and skin. The thief would be conservative, so brown leather was chosen.

While developing the device's physical aspects, I was torn between making it a utilitarian tool, a prop for a specific narrative, or an abstract design object. The tool was rejected on the grounds that the contrast with its non-utilitarian function would be two obvious. The first leather version worked well as a film

Figure 7.10 Object of desire: pacemaker.

prop, but less well as an object in its own right, the object's slightly antique appearance, meant it could be mistaken for a curious antique object, and its poor workmanship made it look too flimsy to be convincing. A second version was designed to explore the expressive possibilities of highly synthetic materials such as upholstered technical fabrics creating a less familiar image for the device. This version was an abstraction of the earlier version and it was covered in thin foam and flesh coloured fabric suggestive of prosthetic limbs and sex toys. The abstract form restrains the overt imagery of the fabric. It yields few clues about its function, other than it is easily wiped clean. Two straps and a plastic ear-nipple offer more suggestions.

Thief of Affections would be presented in a "shoe box," marked with the size of the object (S, M, L, or XL). And the device itself would be designed to last for as long as possible and would not be adjustable. The weak signals picked up by the device would be converted into sounds played through an "ear-nipple" of "prosthetic beige" plastic. The sounds, developed by Jayne Roderick, would range from vaguely masculine to vaguely feminine depending on preference.

Figure 7.11   (see opposite)

The object would be carried, swinging by the side, and would be hoisted to the shoulder for use. The change in position would cause a tilt switch to activate an automatic scan of a range of pacemaker frequencies, locking onto any close signal. Interaction with the object is minimal. Interaction through the object with unsuspecting victims is more important. The device could also be rented for short periods to provide new experiences.

From the start the object was viewed as neither a conventional nonworking mock-up nor a fully working prototype, but as a prop for a series of photographs inspired by a mixture of anthropological and medical photography. I worked with photographer Lubna Hammoud developing a series of scenarios that focus on the psychology of the owner as well as the object, emphasizing the psychosocial narrative possibilities of an electronic object as a role model (figures 7.11–7.12).

The strangeness of the behavioral model embodied in this proposal draws attention to the fact that all electronic products embody models about behaviour and it questions just how distinct our own identity is from those embodied in the electronic objects we use.

Figures 7.11–7.12   Preliminary studies for a portrait of the Thief of Affections.

Tuneable Cities

Tuneable Cities investigates overlapping electromagnetic, urban (and natural) environments. It uses the car as a found environment/object, the product designer's entry point into urbanism. With its built-in radio, telephone, navigator, and even television, the car is already an interface between hertzian and physical space.

The proposal began during a drive across Ireland. While the changing landscape is visible, the car isolates its occupants from corresponding cultural changes and the changing radioscape. The car radio reinforces this by automatically retuning to a station selected at the start of the journey. But if the radio instead tunes automatically to local stations, changes in the landscape would be

Figure 7.13   Sketch of local radio blocks.

matched by changes in culture, interest, and dialect revealing the vernacular qualities of hertzian space. The car would link its occupants to the environment rather than isolating them from it (figure 7.13).

This proposal was connected with an earlier unsubmitted idea for a competition to design a monument for Shepherd's Bush Green roundabout in London. My proposal was an abstract radio sculpture that would produce a radio environment designed to penetrate passing cars. Other transmitters could be located around London, so that when the car radio is tuned to a "public utility" frequency, the car environment becomes a mobile capsule of abstract sound as it moves through the city.

The next stage, in collaboration with Fiona Raby, used a car, wideband radio scanner, scanner directory, and street map to search for interesting overlaps between electromagnetic and urban environments in London. The scanner read interstitial urban radio spaces. British Telecom's research laboratories provided computer models of interactions between cellular phone signals and urban environments that encouraged us to think of these radio spaces as environments.

The scanner was programmed to search for illegal bugging devices as we drove around Park Lane, Gray's Inn, Fleet Street, and Mayfair. It registered possible bugs by momentarily "clicking" on known frequencies that were then marked on a map. We found our first definite bug in Mayfair. By parking close to two edges of a building, we picked up a bugged conversation that we registered on the map.

We then programmed the scanner to receive transmissions from babycoms (baby monitoring intercoms) and drove to suburban Chiswick, a district favored by new families, and were surprised at the extent babycoms were transmitting domestic sounds into the street. We marked our findings on a map and found that on some streets almost half the houses transmitted domestic soundscapes.

We video-recorded some of the areas we had explored, digitised the tapes, and produced short video clips visualizing the radio events as environments and experiences. This stage experimented with different languages of representation. The videos aimed to convey a sense that radio is not only a medium, but is environmental: it can be occupied, extending conventional architectural spaces to blur boundaries between private and public.

Alternative sources of radio were explored. Two further possibilities emerged: mobile fields of abstract sound produced by radio-tagged birds, and natural radio produced by atmospheric events. For *Radio Birds* (figure 7.14), birds (possibly

Figure 7.14   Video still from *Radio Birds* video.

already tagged by scientists) would become radios generating seasonal fields of abstract sound as they migrate. They would either transmit directly to the car or reflect signals transmitted to them from arrays of antennas positioned at ground level or on buildings, which would in turn be transmitted to the car. Alternately, arrays of antennas, functioning as perches, would amplify and transmit the signals to passing cars (figure 7.15).

Public Utility would consist of zones of speed-trap radar. When cars entered these zones their presence would affect the signal they were receiving so that the sound environment of the cars would be directly affected by the cars using the roundabout.

We then used the language generated by the earlier video experiments to present a development of *Public Utility* and *Radio Birds*. We chose two more London sites: Waterloo for *Public Utility* and Trafalgar for *Radio Birds*. Three-dimensional architectural models using synthetic materials and silk-screened

Figure 7.15   Sketch of *Radio Birds* antennas.

Figure 7.16 Sketch of car radio.

maps were also made for each environment to reinforce the shift in emphasis from radio as energy to radio as space.

The final stage of Tuneable Cities shifted attention back to the physical nature of the car radio itself. At first the radio was going to be a clip-on car accessory, an alternative use of the existing car radio slot, or something to do with adhesive patches and tax discs (figure 7.16), but I felt that the essence of this project is the reengineering of a radio as design's potential for subversion lies in the product's function rather than its form.

Ideally, the design of the radio would have been given to an established commercial design practice and aimed at a particular market. As that was beyond the scope of this project, images of existing car radios were modified to show alternative functions organized as legal/illegal, urban/rural, and private/public pairs (figure 7.17). Some presets for Sferics, bugs, babycoms, *Public Utility,* and *Radio Birds* were also included.

The design process behind this proposal acknowledges the electromagnetic spectrum as a social space, where new definitions of private and public are

Figure 7.17   Treated car radio image.

currently being worked out. Illegal bugging devices, and babycoms that unintentionally act as bugs, provide extensions of private spaces into the public realm. Embassies, legal districts, and suburbs are already part of a tuneable city. *Radio Birds* explores relationships between people and an artificial nature mediated by cars, while *Public Utility* draws attention to the disembodied public space shared by transient mobile communities of car drivers.

The Tuneable Cities proposals question which part of a design process needs to be communicated, and how. They take the car as a found environment/object and revisit the city using mass-produced products to explore public and private space, artificial nature, public art, and overlaps between electromagnetic and urban environments. They suggest a role for electronic products as shapers of urban experience.

Faraday Chair: Negative Radio

Science tells us that radio is everywhere at all times. Whenever an electron changes its motion, the disturbance brings about an electromagnetic wave with radio frequency. This means that as you turn the pages of Radiotext(e), you're creating radio waves. So handle this anthology carefully—radio waves never die."
—N. STRAUSS, *RADIOTEXT(E)*

During a project about electronic space, I realized that today all space is electronic, and that the challenge to designers is to create an "empty" space, a space that has not existed for most of this century due to the explosion of uses for the electromagnetic spectrum. This proposal is not concerned with dramatic aesthetic expressions of electronic phenomena but with providing a "conceptual buzz."

My proposed object for presenting a non-electronic, radio-free volume would use a faraday cage to show the ubiquitous nature of radio space and make perceptible the absence of radio. The object would ask: if the inside is empty, what is outside?

My first ideas were for literal faraday cages that, containing something natural like fruit, would imply that natural objects needed to be protected from electromagnetic pollution. But the mesh was easily interpreted as a decorative device and, although the earth lead for the cage conveyed an impression of electrical functionality, it made it look scientific in a quaintly Victorian way. So I sought less technical-looking shield materials and a more powerful subject for enclosure.

The sculptor Jannis Kounellis uses a harsh steel bed frame to represent the absent human. This made me realize that the human body should be what is contained, and that even when unoccupied the object should still refer to the body. In Kounellis's piece, the bed's power as an image stems from both its human scale and the fact that we are born and die on it.

The use of chairs to express prevailing values and ideas about design is well established. Chairs also echo the human body and can communicate new images of man. For example, the bean bag expressed the new informality of the 1960s. The Faraday Chair could provide a new image of the technological person: not of a cyborg fusing with technology, or of a master of technology, but of vulnerability and uncertainty about the long-term effects of the technologies now so enthusiastically embraced.

While exploring ideas for the Faraday Chair, and as a reaction to the awkwardness of the chairs, the daybed suggested an elegant image of an object whose occupant escaped not into a fantastic world of VR but to enjoy the conceptual buzz of a pure electronic radiation-free space. But the calm repose of the daybed seemed too decadent. The awkwardness of the chair conveyed desperation, and its cramped space lent preciousness to the pure space it contained. The final object even if a daybed, should not be too luxurious as it might appear utopian. There should be something not quite right about it. Its origins should appear uncertain.

By this stage I had moved from metal meshes and sheets to tinted acrylics or glass, and to silk-screened conductive inks or conductive films and coatings that were visually transparent but radio-opaque. I rejected fabrics because radio-opaque versions were only available as visually opaque and it was important to highlight the difference between visual and radio transparency.

Figure 7.18 Final version with occupant.

Although the final object was a smaller version of a daybed, requiring the occupant to adopt a fetal position (figure 7.18), I kept the title Faraday Chair to suggest that, once electromagnetic fields are taken into consideration, conventional assumptions about everyday objects need to be reexamined. The final proposal is a compromise between the awkwardness of the chair and the abstract elegance of the daybed. A snorkel mouthpiece attached to a silicone air tube was added to hint at the darker psychological side of the proposal and counter the object's formal elegance (figure 7.19).

I developed a series of photographic scenarios with Lubna Hammoud for this object, too. The emphasis was on portraying the vulnerability of its user. Photography was chosen over video as very specific moments could be constructed and more left to the viewer's imagination.

This proposal draws attention to aesthetic differences between sensual and conceptual approaches to the electronic object. The object is stripped back to its essentials, learning from Electroclimates, the Faraday Chair is less seductive and more difficult to accept.

Figure 7.19 Close-up.

Material Tales

The design proposals described in this chapter function as conceptual test-pieces that, through their strangeness, make visible some of the social and psychological mechanisms that shape aesthetic experiences of everyday life mediated by electronic products. Their apparent unusability creates a heightened sense of "distance." This can be because the objects do not work technically or, preferably, because they are conceptually difficult to assimilate. Through use, or at least by modeling a scenario of use in the mind, the observer discovers new ways of conceptualizing reality. They challenge how we think about extensions to our "selves" in ways that do not simply magnify but, rather, transform our perception and consciousness of our relation to our environment.

They are material tales that allow complex interactions between reality and imagination. Driven by poetry, imagination, and intuition rather than reason and logic, they have their own sense, an alternative to our everyday scientific-industrial one. These are tales about the space between rationality and reality, which in an industrial society have come to be synonymous.

Conclusion

This project set out to develop a design approach for producing conceptual electronic products that encourage complex and meaningful reflection on inhabitation of a ubiquitous, dematerializing, and intelligent environment: a form of social research to integrate critical aesthetic experience with everyday life.

Whereas architecture and fine art often refer to popular culture, industrial design is popular culture. Its language is accessible and appeals to the senses and imagination rather than the intellect. I hope in my approach I have retained the popular appeal of industrial design while using it to seduce the viewer into the world of ideas rather than objects. Industrial design locates its object in a mental space concerned with identity, desire, and fantasy and shaped by media such as advertising and television. Again, I hope this remains intact but is subverted to challenge the aesthetic values of both consumers and designers.

One result of this research is a toolbox of concepts and ideas for developing and communicating design proposals that explore fundamental issues about how we live among electronic objects. The most important elements of this approach are: going beyond optimization to explore critical and aesthetic roles for electronic products; using estrangement to open the space between people and electronic products to discussion and criticism; designing alternative functions to draw attention to legal, cultural, and social rules; exploiting the unique narrative possibilities offered by electronic products; raising awareness of the electromagnetic qualities of our environment; and developing forms of engagement that avoid being didactic and utopian. The emphasis on behavior as narrative experience stimulated by the design of new functions links these proposals more to the complicated pleasure of literature and film than to sculpture—the traditional reference for industrial design.

But perhaps the most important conclusion is that design is existential and cannot ignore its complicated relationship to people and their mental lives. This project offers one possible approach to allow some of the more complicated and difficult aspects of our relationship to electronic objects to be reflected in future designs.

Notes

Introduction

1. The mainstream view of industrial design serving the narrow commercial interests of industry as opposed to a more general social role for design: developing tools for living.

2. The adventurous spirit and humanist vision of Italian design research from the 1960s, 1970s, and early 1980s has been an inspiration throughout this project. The design approach developed for electronic objects in this book is indebted to their research.

3. "I feel myself that the writer's role, his authority and licence to act, have changed radically. I feel that, in a sense the writer knows nothing any longer. He has no moral stance. He offers the reader the contents of his own head, he offers a set of options and imaginative alternatives. His role is that of the scientist, whether on safari or in his laboratory, faced with a completely unknown terrain or subject. All he can do is to devise various hypotheses and test them against the facts." J. G. Ballard, *Crash,* 9.

4. In his introduction to *Einstein's Monsters,* Martin Amis (1987) writes that the purpose of the stories is "no purpose at all—except, I suppose, to give pleasure, various kinds of complicated pleasure" (ix).

1 The Electronic as Post-optimal Object

1. The idea of a "post-optimal" object arose from a general observation made by Marco Susani during a workshop at the Netherlands Design Institute in 1996. His suggestion, that products in general had reached an optimal level, and that designers should turn

their attention toward cities and urban environments that were far from optimal, suggested to me that there must be something beyond the optimal. The idea of a post-optimal object could rescue design objects concerned with ideas, from the clutches of the art world, while maintaining a relationship to design.

2. For more on the analysis of objects as commodities, see A. Appadurai, "Introduction: Commodities and the Politics of Value."

3. "Scientific theories have as their aim or goal successful manipulation of the external world; they have instrumental use. If correct, they enable the agents who have mastered them to cope effectively with the environment and thus pursue their chosen ends successfully. Critical theories aim at emancipation and enlightenment, at making agents aware of hidden coercion, thereby freeing them from that coercion and putting them in a position to determine where their true interests lie." R. Geuss, *The Idea of a Critical Theory,* 55–56.

4. "Electrons are the smallest of these particles and each one carries the smallest amount of negative electricity. In most materials and especially in good insulators like glass or plastics electrons are held firmly in place by heavier, positively charged protons. Some materials, mainly metals, contain electrons that have enough energy to move about allowing the electrons to transport electricity from one place to another, making metals good conductors. When metals are heated, their electrons are given even more energy, sometimes causing them to completely break free from the metal. If they are freed in a vacuum where there is nothing for them to collide with, they can be guided by electricity or magnetism to form controlled electric current. This is the basic principle of the valve and the cathode ray tube." R. Bridgman, *Electronics,* 26.

5. Thin films of polymer material in which particles of doped zinc sulphide have been absorbed, and which, inserted into a sandwich of other protective materials and subjected to alternating current, emit uniform luminosity across their entire surface.

6. These transform mechanical impulses into electric impulses or vice versa and are widely used to produce sensors, actuators, microphones, and loudspeakers.

7. A group of theoretical and design experiments carried out by Andrea Branzi and Clino Trini Castelli during the mid-1970s. Emphasis was placed on "experiences of space that are not directly assimilable to the constituent qualities of an environment or an object, but are linked instead with the physical perception of space, i.e. with its 'bodily' consumption. In this way new attention was paid to the user's real sensitivity of perception, bound up more closely with the direct consumption of soft structures than with

the grasping of an architectural composition and its sophisticated allegories of form."
A. Branzi, *The Hot House,* 97–98.

2   (In)human Factors

1. For an excellent critique of product semantics, see A. Richardson, "The Death of the Designer."

2. For a summary of John Dewey's views on aesthetic experience in terms of recognition and perception, see chapter 4.

3. Irie and Fujihata's approaches superficially resemble that of John Frazer who has been involved in computer-generated form and structure since the early 1960s. Like many explorations of autogenerative models, especially in the field of artificial life, Frazer's inventions rarely move beyond the screen into physical space, although their formation often responds to data, such as environmental conditions, from sources outside the computer. See J. Frazer, *Themes VII: An Evolutionary Architecture.*

3   Para-functionality: The Aesthetics of Use

1. This also suggests a way of establishing an architectural role for the object in the sense of Bernard Tschumi's "there is no space without event, no architecture without programme; the meaning of architecture, its social relevance and its formal invention, cannot be dissociated from the events that 'happen' in it." B. Tschumi, "The Discourse of Events," 17.

2. "The Japanese word 'Chindogu' literally means an odd or distorted tool—a faithful representation of a plan that doesn't quite cut the mustard . . . they are products that we believe we want—if not need—the minute we see them. They are gadgets that promise to give us something, and it is only at second or third glance that we realise that their gift is undone by that which they take away." K. Kawakami, *101 Unuseless Japanese Inventions,* 6–7.

3. Thackara, *Design after Modernism,* 22.

4. I refer to those cultural mechanisms that marginalize alternatives to the present, even when economically and technically feasible, as utopian and "unrealistic."

5. This project is of personal interest to me because a similar project, the *Noiseman* (1989) marked my first experience of designing in a critical way while working for a

consumer electronic company. My *Noiseman,* less self-consciously political than Wodiczko's work, focused on the role of mass production and on creating new aesthetic experiences of urban spaces.

6. A wideband radio of the kind that, tuned into a mobile phone conversation between Princess Diana and a lover, recorded the "Squidgy tapes," extracts of which were published in the tabloids in early 1990s.

4   Psychosocial Narratives

1. "An urban legend appears mysteriously and spreads spontaneously in varying forms. Contains elements of humour or horror (the horror often 'punishes' someone who flouts society's conventions). Makes good storytelling. Does NOT have to be false, although most are. ULs often have a basis in fact, but it's their life after-the-fact (particularly in reference to the second and third points) that gives them particular interest." From FAQ alt.urban legends.

2. "Its pursuance of current lifestyle as the sustaining parallel to the design of cities forms the basis of its spirit and optimism." Opening comment in each issue of *Narrative Architecture Today.*

3. "So technology might be the passport back to a world where concepts and images are more important than actuality. In the manner of reclusants and hermits of yore, by his handling of soft and hardware, modern man can be a dreamer in meta-mataphysical [sic] space. If however he is unready to discard the reality of this earth in favour of a manufactured romance, at least his technologically induced, media fed sensibilities can be better married to the outside world. If he knows fictional distortions in time and in space, it becomes easier to settle in real places that are learning to use an apposite technique of expressiveness." Beevor, "Between Here and Now," 6.

4. "Unfortunately, Mead's original meaning of the term 'role model' has become narrowed, so that now social scientists tend to emphasize the behavioral patterns of an actual person as constituting a 'role model,' leaving out or omitting the fact that Mead includes 'any object' or 'set of objects' as having this power as well." Csikszentmihalyi and Rochberg-Halton, *The Meaning of Things,* 51.

5   Real Fiction

1. Although it is not an electronic product, when Krzysztof Wodiczko's *Poliscar* was exhibited at the Josh Baer Gallery, he was criticized for "'revelling in the mechanics' of his

warlike contraption when he might be more usefully have been working out how it could 'transform current social realities.'" Yet that is exactly what the artist was concerned with. P. Wright, "The Poliscar: Not a Tank but a War Machine for People without Apartments."

2. I use the term "genotype" as an alternative to "prototype" to shift importance away from whether or not a conceptual design technically works, to the ideas it represents.

6 Hertzian Space

1. Not only radio broadcast frequency but general Radio Frequency: 10 kilohertz–0.1 terahertz.

2. Michele Bertomen's (1991) *Transmission Towers on the L.I.E. (Long Island Expressway)* occasionally mentions electromagnetic space but is primarily a typology of the physical structures that support it. The cultural history of the radio-frequency part of the electromagnetic spectrum and its many uses are well documented in Augaitis and Lander's (1994) *Radio Rethink* and Strauss's (1993) *Radiotext(e).* More philosophical and political aspects have been dealt with by Virilio (1991, 1995) in *The Lost Dimension* and *The Art of the Motor,* and Manual DeLanda's (1991) *War in the Age of Intelligent Machines.*

3. M. Sorkin, "Architecture Rising: Lebbeus Woods' Paris Project."

4. L. Kurgan, "You Are Here: Information Drift."

5. "Download the data to a computer from the receiver, then download from a local base station its (scattered) readings for the same time period (Reference). Correct your readings and reduce the drift (Correction), average the points, and learn where you were— within a few meters (Average). In the computer, the satellites draw the points for you; as the readings become more precise the points grow to fill the screen." Ibid., 18.

6. M. L. Heivly and M. Reed, "The Space between the Real and the Imagined: Microwave Sculpture in Deep Space."

7. Some fish use the three-dimensional equivalent of radar to guide themselves through dark waters by positioning themselves in each others' fields, possibly communicating through voltage patterns and rates of pulsing, a sort of electrical whisper. L. Milne and M. Milne, *The Senses of Animals and Men,* 109–120.

8. For more on early radio detectors, see V. J. Philips, *Early Radio Wave Detectors.*

9. Although I have not encountered this term in any of Ito's writings, he used it often during the design of the *Dreams* room for the Victoria and Albert Museum in London.

10. Five "Interactive Terminals" were designed by myself and Fiona Raby for the exhibition.

Bibliography

Abalos, I., and J. Herreros. 1995. "Toyo Ito: Light Time." In *El Croquis,* vol. 71.

Aldersey-Williams, H. 1990. *The New Cranbrook Design Discourse.* New York: Rizzoli.

Amis, M. 1987. *Einstein's Monsters.* London: Penguin.

Appadurai, A. 1986. "Introduction: Commodities and the Politics of Value." In *The Social Life of Things: Commodities in Cultural Perspective,* ed. A. Appadurai. Cambridge: Cambridge University Press.

Apple Computer. 1992. "Knowledge Navigator." *CHI '92 Special Video Program Issue 79.* Siggraph Video Review (video, 1987).

Augaitis, D., and D. Lander, eds. 1994. *Radio Rethink.* Banff: Walter Philips Gallery.

Bachelard, G. 1969. *The Poetics of Space.* Boston: Beacon.

Baker, N. 1986. *The Mezzanine.* Cambridge: Granta.

Baker, N. 1990. *Room Temperature.* London: Granta.

Baker, N. 1992. *Vox.* London: Granta.

Ballard, J. G. 1990. *Crash.* London: Paladin.

Banham, R. 1981. *Design by Choice.* London: Academy.

Banham, R. 1960. *Theory and Design in the First Machine Age.* London: Architectural Press.

Barthes, R. 1989. *Mythologies.* London: Paladin.

Baudrillard, Jean. 1981. *For a Critique of the Political Economy of the Sign.* St. Louis: Telos.

Baudrillard, Jean. 1983. "The Ecstasy of Communication." In *The Anti-Aesthetic,* ed. H. Foster. Seattle: Bay Press.

Baudrillard, Jean. 1988a. *The Ecstasy of Communication.* New York: Semiotext(e).

Baudrillard, Jean. 1988b. "The System of Objects." In *Design after Modernism,* ed. John Thackara. London: Thames & Hudson.

Baudrillard, Jean. 1990. *The Revenge of the Crystal.* London: Pluto.

Baudrillard, Jean. 1993. "Xerox and Infinity." In *The Transparency of Evil.* London: Verso.

Baudrillard, Jean. 1996. "The Metafunctional and Dysfunctional System: Gadgets and Robots." In *The System of Objects.* London: Verso.

Becker, C. 1994. "Herbert Marcuse and the Subversive Potential of Art." In *The Subversive Imagination,* ed. C. Becker. New York and London: Routledge.

Beevor, C. 1983. "Between Here and Now." In *Narrative Architecture Today (NATO).* London: Architectural Association.

Bender, G., and T. Druckrey, eds. 1994. *Culture on the Brink.* Seattle: Bay Press.

Bertomen, M. 1991. *Transmission Towers on the L.I.E. (Long Island Expressway).* New York: Princeton Architectural Press.

Birnbaum, A. 1991. "Fexy Facts." *Mediamatic* 5(4).

Bolter, D. J. 1993. *Turing's Man.* London: Penguin.

Bolz, N. 1992. "The Meaning of Surface." In Netherlands Design Institute, *Doors of Perception 1 DopRom* (CD-ROM). Amsterdam: NDI, 1994.

Bouman, O. 1996. *Real Space in Quick Times: Architecture and Digitisation.* Rotterdam: Nai.

Branzi, A. 1984. *The Hot House.* London: Thames & Hudson.

Branzi, A. 1992. *Luoghi Andrea Branzi: The Complete Works.* London: Thames & Hudson.

Branzi, A. 1995. "Crisis and Growth: Industrial Design and the Creative Tradition." In S. Marzano, *Television at the Crossroads.* London: Academy Editions.

Bridgman, R. 1993. *Electronics.* London: Dorling Kindersley.

Burns, C., et al. 1994. "Actors, Hairdos and Videotape—Informance Design." In *Human Factors in Computing Systems CHI '94 Conference Proceedings.* New York: ACM Press.

Calas, H., and N. Calas. 1984. *Takis.* Paris: Éditions Galilée.

Ceppi, G. 1991. "Playing with Technology." In *Modo* 136 (1991).

Chant, C., ed. 1989. *Science, Technology and Everyday Life 1870–1950.* London: Routledge.

Chapman, G. 1994. "Making Sense Out of Nonsense." In *Culture on the Brink,* ed. G. Bender and T. Druckrey. Seattle: Bay Press.

Choe, W. 1989. *Toward an Aesthetic Criticism of Technology.* New York: Peter Lang.

Clarke, K. 1967. *Positioning in Radiography.* London: Heinemann.

Coleman, R. 1993–1994. *Design Research for Our Future Selves, Vol. 2.* London: Royal College of Art.

Csikszentmihalyi, M., and E. Rochberg-Halton. 1981. *The Meaning of Things.* Cambridge: Cambridge University Press.

Daly, N., and B. Lyons. 1991. "The Lost Civilization of Llhuros: The First Multimedia Exhibition in the Genre of Archaeological Fiction." *Leonardo* 24(3).

DeLanda, M. 1991. *War in the Age of Intelligent Machines.* New York: Zone.

De Lucchi, M., and S. Marzano. 1994. *Workshop.* Eindhoven: Phillips/Olivetti.

Dewey, J. 1958. *Art as Experience.* New York: Minton, Balch.

Dike, C. 1983. *Cane Curiosa: From Gun to Gadget.* Paris: Éditions de L'Amateur.

Dollens, D. 1991. "Toward Ninja Architecture/hyperarchitexture." *Off Ramp* 1(4).

Dormer, P. 1990. *The Meanings of Modern Design.* London: Thames & Hudson.

Dunne, A., and F. Raby. 1995. "Fields and Thresholds." In *Architects in Cyberspace,* ed. Neil Spiller. Architectural Design Profile No. 118. London: VCH.

Dunne, A., and W. Gaver. 1996. "The Pillow: Artist-Designers in the Digital Age." In *Human Factors in Computing Systems CHI '96 Conference Proceedings.* Reading, MA: Addison-Wesley.

Everett, J. D. 1904. *Deschanel's Natural Philosophy, Part 3.*

Feibleman, J. K. 1982. *Technology and Reality.* The Hague and London: Nijhoff.

Form Finlandia. 1987. Promotional leaflet for design competition.

Forty, A. 1986. *Objects of Desire.* London: Thames & Hudson.

Foucault, M. 1970. *The Order of Things.* London and New York: Tavistock/Routledge.

Frampton, K., and S. Kolbowski, eds. 1981. *Idea as Model.* New York: Rizzoli.

Frazer, J. 1995. *Themes VII: An Evolutionary Architecture.* London: Architectural Association.

Freud, S. 1966. *Jokes and Their Relation to the Unconscious.* London: Routledge.

Fujihata, M. 1991. *Forbidden Fruits.* Tokyo: Libro Port.

Garner, P. 1983. *Philip Garner's Better Living Catalogue.* London: Sidgwick & Jackson.

Garner, P. 1985. *Utopia or Bust.* London: Plexus.

Geuss, R. 1981. *The Idea of a Critical Theory.* Cambridge: Cambridge University Press.

Grassmuck, V. 1991. "Otaku." *Mediamatic* 5(4).

Griffiths, P. 1986. *Modern Music.* London: Thames & Hudson.

Hawkes, T. 1992. *Structuralism and Semiotics.* London: Routledge.

Heivly, M. L., and M. Reed. 1992. "The Space between the Real and the Imagined: Microwave Sculpture in Deep Space." *Leonardo* 25(1): 17–21.

Hubert, C. 1981. "The Ruins of Representation." In *Idea as Model,* ed. K. Frampton and S. Kolbowski. New York: Rizzoli.

Irie, K. 1988. "Computer Crash by Design." *Telescope* (May/June).

Irie, K. 1993. "AA Intermediate Unit 3: Objects in the Landscape." *AA Files* 25.

Ito, T. 1995."Architecture in a Simulated City." In *El Croquis,* vol. 71.

Johnson, P. 1984. *Marxist Aesthetics: The Foundations within Everyday Life for an Emancipated Consciousness.* London: Routledge & Kegan Paul.

Kawakami, K. 1995. *101 Unuseless Japanese Inventions.* London: Harper Collins.

King, S. N.d. "Scanner." In *Trance Europe Express 2,* ed. H. Mead. London: BMG (UK) Ltd.

Kipnis, J. 1991. "ATM Competition." *Off Ramp* 1(4).

Klüver, B. 1994. "Artists, Engineers, and Collaboration." In *Culture on the Brink,* ed. G. Bender and T. Druckrey. Seattle: Bay Press.

Krippendorff, K., and R. Butter, eds. 1989. *Design Issues* 5(2).

Kunstflug Formgestaltung. 1991. "Design without an Object." *Ottagono* 100.

Kurgan, L. 1995. "You Are Here: Information Drift." *Assemblage* 25: 14–43.

Lander, D. 1994. "Radio and Beans." In *Radio Rethink,* ed. D. Augaitis and D. Lander. Banff: Walter Philips Gallery.

Lee, Ray, and Harry Dawes. 1996. Photocopy distributed at performance of *In the Ether.*

Mann, S. 1992. "DopplerDanse: Some Novel Applications of Radar." *Leonardo* 25(1).

Manzini, E. 1986. *The Material of Invention.* Milan: Arcadia.

Manzini, E. 1992. "Prometheus of the Everyday: The Ecology of the Artificial and the Designer's Responsibility." *Design Issues* 9(1).

Manzini, E. 1994. "Design, Environment and Social Quality: From 'Existenzminimum' to 'Quality Maximum.'" *Design Issues* 10(1): 37–42.

Manzini, E. 1994. "Design Research for a Sustainable Development." Paper delivered at Design/Research Conference, Royal College of Art, May.

Manzini, E., and M. Susani, eds. 1995. *The Solid Side.* Netherlands: V + K.

Margolin, V., and R. Buchanan, eds. 1995. *The Idea of Design.* Cambridge, MA: MIT Press.

Marzano, S. 1993. "Flying Over Las Vegas." *European Design Centre (EDC) News,* September.

Marzano, S. 1995. *Television at the Crossroads.* London: Academy Editions.

McLuhan, M. 1970. *Counterblast.* London: Rapp & Whiting.

Mead, H., ed. N.d. *Trance Europe Express 2.* London: BMG (UK) Ltd.

Mead, H. G. 1934. *Mind, Self, and Society from the Standpoint of a Social Behaviourist.* Chicago: University of Chicago.

Mideke, M. 1991. *A Whistler Hunter's Guide.* Published by the author.

Miller, D. 1987. *Material Culture and Mass Consumption.* Oxford: Blackwell.

Milne, L., and M. Milne. 1963. *The Senses of Animals and Men.* London: Andre Deutsch.

Moles, A. 1995. "Design Immateriality: What of It in a Post Industrial Society?" In *The Idea of Design,* ed. V. Margolin and R. Buchanan. Cambridge, MA: MIT Press.

Moore, J. R. 1989. "Everyday Life and the Dynamics of Technological Change." In *Science, Technology and Everyday Life 1870–1950,* ed. C. Chant. London: Routledge.

Netherlands Design Institute. 1994. *Doors of Perception 1 DopRom* (CD-ROM). Amsterdam: NDI.

Nicholson, B. 1990. *Appliance House.* Cambridge, MA: Chicago Institute for Architecture and Urbanism.

Norman, D. 1988. *The Psychology of Everyday Things.* New York: Basic Books.

Norman, D. 1993. *Things That Make Us Smart: Defending Human Attributes in the Age of the Machine.* Reading, MA: Addison-Wesley.

Philips, V. J. 1980. *Early Radio Wave Detectors.* London: Institution of Electrical Engineers.

Philips Corporate Design. 1996. *Vision of the Future.* Bussum, Netherlands: V+K.

Porush, D. 1985. *The Soft Machine.* New York: Methuen.

Popper, F. 1968. *Origins and Development of Kinetic Art.* London: Studio Vista.

Popper, F. 1993. *Art of the Electronic Age.* London: Thames & Hudson.

Radice, B. 1985. *Memphis.* London: Thames & Hudson.

Richardson, A. 1993. "The Death of The Designer." *Design Issues* 9(2).

Robinson, R. 1994. "Review of *Things That Make Us Smart.*" *Design Issues* 10(1).

Rochberg-Halton, E. 1979a. "The Meaning of Personal Art Objects." In *Social Research and Otium Publications,* ed. J. Zuzanek. Waterloo, Ontario: Otium Publications.

Rochberg-Halton, E. 1979b. "Cultural Signs and Urban Adaptation: The Meaning of Cherished Household Possessions." Unpublished Ph.D. diss., University of Chicago.

Ross, A. 1994. "The New Smartness." In *Culture on the Brink,* ed. G. Bender and T. Druckrey. Seattle: Bay Press.

Rubinstein, M. R. 1993. "Philippe Ramette." *Art Forum* (May).

Schafer, M. 1993. "Radical Radio." In *Radiotext(e),* ed. N. Strauss. New York: Semiotext(e).

Shinohara, K. 1982. *Kazuo Shinohara.* New York: Rizzoli.

Silverstone, R., and E. Hirsch. 1992. *Consuming Technologies.* London: Routledge.

Smith, C. W., and S. Best. 1989. *Electromagnetic Man.* London: Dent.

Sorkin, M. 1990. "Architecture Rising: Lebbeus Woods' Aerial Paris Project." *Didalos* (December).

Sorkin, M. 1992. *Variations on a Theme Park.* New York: The Noonday Press.

Sparke, P. 1982. *Ettore Sottsass Jnr.* London: Design Council.

Speckley Pittman PR. 1994. *Mobile Moments.* London: Telecom Securicor Cellular Radio Ltd.

Strauss, N., ed. 1993. *Radiotext(e).* New York: Semiotext(e).

Sturrock, J. 1986. *Structuralism.* London: Paladin.

Surveillance Technology Group. N.d. *Covert Audio Intercept* (company brochure). Port Chester: STG.

Susani, M. 1992. "The Fourth Kingdom." *Ottagono* 105.

Susani, M. 1994. "The Sensible Home." Paper delivered at Doors of Perception 1 Conference, Amsterdam, November.

Tanizaki, J. 1991. *In Praise of Shadows.* London: Jonathan Cape.

Thackara, J. 1986a. "The Cacophonous Sounds of Technology." *AA Files* 11.

Thackara, J. 1986b. *New British Design.* London: Thames & Hudson.

Thackara, J. 1988. *Design after Modernism.* London: Thames & Hudson.

Tschumi, B. 1983. "The Discourse of Events." In *Themes III.* London: Architectural Association.

Tschumi, B. 1994a. *Architecture and Disjunction.* Cambridge, MA: MIT Press.

Tschumi, B. 1994b. *Event-Cities.* Cambridge, MA: MIT Press.

Van Weelden, D. 1992. "Machine Voices." In *Mediamatic* 6(4).

Vidler, A. 1992. *The Architectural Uncanny.* Cambridge: MIT Press.

Virilio, V. 1991. *The Lost Dimension.* New York: Semiotext(e).

Virilio, V. 1995. *The Art of the Motor.* Minneapolis: University of Minnesota Press.

Waites, B. 1989. "Everyday Life and the Dynamics of Technological Change." In *Science, Technology and Everyday Life 1870–1950,* ed. C. Chant. London: Routledge.

Weibel, P. 1994. *Intelligent Ambience.* Promotional leaflet for Ars Electronica. Linz: ORF.

Weller, P., W. Mackay, and R. Gold, eds. 1993. "Computer-Augmented Environments: Back to the Real World." In *Communications of the ACM* 36(7).

Wodiczko, K. 1992. *Instruments, Projections, Vehicles.* Barcelona: Fundació Antoni Tàpies.

Woods, L. 1985. *Origins.* London: Architectural Association.

Wright, P. 1992. "The Poliscar: Not a Tank but a War Machine for People without Apartments." In *Instruments, Projections, Vehicles,* ed. K. Wodiczko. Barcelona: Fundació Antoni Tàpies.

Zimmerman et al. 1995. "Applying Electric Field Sensing to Human Computer Interfaces" In *Human Factors in Computing Systems CHI '95 Conference Proceedings.* Reading, MA: Addison-Wesley.

Illustration Credits

1.1   Daniel Weil, Pentagram Design, *Radio in a Bag* (1983).

1.2   The first working transistor (1947). Replica. Courtesy Science Museum/ Science and Society Picture Library.

1.3   Daniel Weil, Pentagram Design, *Clock* (1983).

1.4   Andrea Branzi, *Leaf Electroluminescent Lamp* (1988). Photograph: Santi Caleca.

1.5   Kunstflug Formgestaltung, *The Electronic Room: Programmable Appearances— Surfaces, Appliances, Comfort.*

1.6   Kunstflug Formgestaltung, *Electronic Hand Calculator.* Photograph: Walter Vogel.

1.7   Fiona Raby, *Balcony* (1995).

1.8   Robert Rauschenberg, *Oracle* (1965). Copyright Robert Rauschenberg/ VAGA, New York/DACS, London 1999.

2.1   Ettore Sottsass ELEA 9003 (1959). Photograph: Sala Dino. Courtesy Archivio Storico Olivetti, Ivrea, Italy.

2.2   Marco Zanuso and Richard Sapper, Brion Vega television (1969). Photograph: Aldo Ballo.

2.3	Matthew Archer, miniature computer (1985).

2.3   Matthew Archer, miniature computer (1985).

2.4   Winfried Scheuer, *Fan Heater* (1981).

2.5   Lisa Krohn, with assistance from Tucker Viemeister, answerphone (1987).

2.6   Mario Trimarchi and Marco Susani, *New Tools (for the kitchen)* (1992). Cultural and technical support: Stefano Marzano, Philips Corporate Industrial Design. Exhibited at Il Giardino Delle Cose in the Triennale Di Milano. Photograph: Santi Caleca.

2.7   Stelarc, *Third Hand* (1976–1981).

2.8   Peter Stathis, *Satori TV* (1988).

2.9   Alan Rath, *C-Clamp* (1992). Courtesy Haines Gallery, San Francisco, CA.

2.10   Daniel Weil, Pentagram Design, *Four Boxes and One Radio* (1983).

2.11   Daniel Weil, Pentagram Design, *Small Door* (1986).

2.12   Kei'ichi Irie, *Lascaux Chair* (1988).

2.13   Masaki Fujihata, *Forbidden Fruits* (1991).

3.1   Jack Kevorkian, *Suicide Machine.* Photograph: Jean-Louis Atlan/Sygma.

3.2   Drinking cane. Courtesy Patrick D. Gutknecht.

3.3   Table cane. Courtesy Patrick D. Gutknecht.

3.4   Seat cane. Courtesy Patrick D. Gutknecht.

3.5   1940s steering device for preventing the destruction of marks on a steering wheel.

3.6   Kenji Kawakami, *The Diver-Drier.*

3.7   Gaetano Pesce, furniture for Cassina (1960s).

3.8   Denis Santachiara, *Portale* (1989).

3.9   Philip Garner, *Alienature* (1985).

3.10   Paul Klee, *The Twittering Machine* (1922). Copyright DACS 1999.

3.11   Panamarenko, *Voyage to the Stars* (1979).

3.12   Philippe Ramette, *Object with Which to See the World in Detail* (1990).

3.13   Andrea Zittel, *A to Z Comfort Unit II* (1994). Courtesy Andrea Rosen Gallery, New York. Photograph: Peter Muscato.

3.14   Ken Kaplan and Ted Krueger, *Bureau-dicto,* 1989. Reprinted with permission from *Pamphlet Architecture 14: Mosquitoes* by Ken Kaplan and Ted Krueger (New York: The Princeton Architectural Press, 1993).

3.15   Ken Kaplan and Ted Krueger with Chris Scholtz, *Crib-batic.*

3.16   Diller + Scofidio, *Para-Site* (1989).

3.17   Krzysztof Wodiczko, *Personal Instrument* (1969). Courtesy Galerie Lelong, New York.

3.18   Krzysztof Wodiczko, *Homeless Vehicles* (1988–1989). Courtesy Galerie Lelong, New York.

3.19   AOR8200 radio scanner. AOR LTD, Japan.

3.20   Dr. Gauss EMF detector. Huntar Company Inc.

4.1   Fax Sharaku™, Fuji-Xerox.

4.2   Architekturbüro Bolles + Wilson, *Ninja House—A Comfortable House in the Electronic Metropolis* (1988). First prize Shinkenshiku Competition.

4.3   Ben Nicholson, *Appliance House* (1990). Cambridge, MA: Chicago Institute for Architecture and Urbanism.

4.4	Toyo Ito & Associates, Architects, *Dwelling for Tokyo Nomad Woman* (1985). Photograph: Tomio Ohashi.
4.5	*Truth Phone*™, manufactured by CCS International Ltd.
5.1	Anthony Dunne, *Monitor as Material* (1996).
5.2	Krzysztof Wodiczko, *Alien Staff* (1992). Courtesy Galerie Lelong, New York.
5.3	Gregory Green, *Nuclear Device #2. 15 kilotons, plutonium 239* (1995).
5.4	Andrea Branzi, *Ginger* (1980).
5.5	Michele De Lucchi, *Appliances* (1979).
5.6	Philips Design and Olivetti, *Communicator* (from Workshop, 1994).
5.7	Philips Design and Olivetti, *Group Tool* (from Workshop, 1994).
5.8	Philip Garner, *Skeet Bowling* (1985).
5.9	Philip Ramette, *Object for Seeing Oneself Watching* (1990).
6.1	Electromagnetic Spectrum showing the Radio Frequency Spectrum.
6.2	Ken Krayer, *Telephone* (1988).
6.3	Radio propagation map. Courtesy British Telecom Research Laboratories.
6.4	Stephen J. Birkill, satellite footprint map (1992). Copyright Stephen J. Birkill, Electronics Design House Real-World Technology Ltd.
6.5	X-ray pose. Lefeuvre's "physiological" receiver.
6.6	Toyo Ito & Associates, Architects, *Tower of Winds* (1986).
6.7	Toyo Ito & Associates, Architects, *Dreams Room* (1991–1992). Photograph: Naoya Hatakeyama.

Index